A Text Book of Environmental Science

(BASED ON UGC PRESCRIBED CORE MODULE SYLLABUS FOR THE UNDER GRADUATE STUDENTS OF ALL DICIPLINES)

A TEXT BOOK OF ENVIRONMENTAL SCIENCE

(BASED ON UGC PRESCRIBED CORE MODULE SYLLABUS FOR THE UNDER GRADUATE STUDENTS OF ALL DICIPLINES)

P. C. JOSHI

Deptt. Of Zoology and Environmental Sciences
Gurukula Kangri University, Hardwar

&

NAMITA JOSHI

Deptt. Of Environmental Sciences
Kanya Gurukula Mahavidyalaya
Gurukula Kangri University, Hardwar

A.P.H. PUBLISHING CORPORATION

4435-36/7, ANSARI ROAD, DARYA GANJ
NEW DELHI-110 002

Published by
S.B. Nangia
A P H Publishing Corporation
4435-36/7, Ansari Road, Daryaganj
New Delhi 110002
Ph.: 23274050
E-mail : aphbooks@gmail.com

2026

Rs. 1795/-

Printed at
Balaji Offset
Navin Shahdara, Delhi 110032

PREFACE

This book has been organized according to the UGC prescribed syllabus of Environmental Studies for under graduate students of all disciplines. The entire book has been divided into eight units as has been given in the UGC syllabus, and each unit has further divided into different chapters.

Unit one has been discussed in chapter one of the book, which deals with the multidisciplinary nature of environment, various definitions of environment, branches of ecology, and scope and importance of ecology. Unit two has been divided into 6 chapters. Chapters two deals with definitions of natural resources, renewable and non-renewable resources and their conservation. Chapter 3 to 7 deal with forest resources, water resources, mineral resources, energy and food resources and land resources, their use and over-exploitation and methods of their conservation. Unit three has two chapters, chapter eight deals about the concept of ecosystem its structure and function, producers, consumers and decomposers. Energy flow in the ecosystem and different types of ecosystems their structure and functions have been discussed in this chapter. Chapter nine is about the ecological succession and its importance. Unit four has one chapter, chapter ten, which is devoted to the subject of biodiversity and its conservation. Unit five is discussed in chapter eleven and twelve. Chapter 11 deals with different types of pollutions their causes, effects and control. It also deals with the role of individuals in prevention of pollution and pollution case studies. Chapter twelve covers natural disasters and their management particularly with reference to India. Unit six has been divided into two chapters. Chapter 13 deals with various social issues and the environment, while chapter 14 deals with various Acts , which have been discussed very briefly. Unit seven is discussed in chapter 15 which included issues related with human population and the environment. Unit 8 of the syllabus is covered under chapter 16 of this book. A guideline has been prepared for the students to perform their field studies and to observe various assets in the environment and study common plants, insects and birds.

Authors express their gratitude towards *Prof. B.D. Joshi,* Principal College of Science, Gurukula Kangri University, Haridwar for his guidance and encouragement during writing of this book. Authors are also highly grateful to all the scientists, writers, researchers and academicians whose work has been cited and or made use of in different chapters of the book. Finally we are grateful to the publishers for their interest in the subject and publishing the book.

P. C. Joshi

Namita Joshi

CONTENTS

CHAPTER-I

Nature of Environmental Studies: A Multidisiplinary Subject

INTRODUCTION

As all living organisms, whether plants or animals are in one or another way inter-dependent on each other, so any such study about the inter-relation ship of biotic (life bearing) and abiotic (life less) factors or components of our environment forms the basis of the ecology.

What we study presently in this branch of biology was well known to our ancient people. For example what we study under the head of physical factors today viz. air, water, soil, light and temperature were considered as the five fundamental components of life by our ancient saints and seers (Rishies). They called them prithvi (The earth –soil), jal (the water), vayu (the air), tej (the light and heat) and akas (the space), from which the sound arose, as described in our 4000-5000 years old vedic literature.

Theophrastus (370-250 B.C.) must be regarded as among the most earlier persons who described the inter-relationships of plants and environmental factors, in a pre-scientific manner and he tried to find out the causes of various modes of life, under varying natural conditions. The 17th and 18th centuries are often regarded as the era of modern scientific dawn, for what ever progress in the field of science and technology we find today, was founded during these two hundred years.

Linnaeus (1707-1798) gave us systematics of biology, Buffon (1707-1788), Malthus (1798) and Doubleday (1841) studied in details about the population characters. However, it was Saint Hilairé (1859), a mineralogist and Professor of Zoology in the National History Museum in Paris, who coined the term 'Ethology', to describe the study of relationship between the living organisms and their environment. But unfortunately, his work did not get due attention. Reiter (1867), introduced the term "Oekologie" derived from two Greek words , "oikos" meaning house or place to live and "ology" means to study, However, quite interestingly, the credit goes to Enrst Haeckel (1869), who gave us the term Ecology, for the study of natural science and its inter-relation with living organisms. Finally, by the end of 19th century and at the dawn of 20th century Ecology had established itself as a full subject.

DEFINITIONS OF ECOLOGY

The brief account given above, gives us an idea of what ecology may be. Even then it is essential to find out the simplest, scientific, effective and correct academic definition for the subject.

Haeckel (1869) said "Ecology is the total relationship of animals to both its organic and inorganic environment."

Clarke (1954), states that "The study of the inter-relations of plants and animals with their environment constitutes the science of ecology." He agrees with the economical aspects of ecology-given in the form of a definition by Wells, Huxley and Wells (1939), which defines it as "Ecology is really an extension of economics to the whole world of life."

Woodburry (1955) has defined ecology as "A science which investigates organisms in relation to their environment : a philosophy in which the world of life is interpreted in terms of natural process".

Odum (1963), another great ecological writer of our times goes further to define ecology as "***The study of the relations of organisms or groups of organisms to their environment***" or ***the science of interrelations between living organization and their environment***. Odum may be cited, in short, as saying "***The ecology is the study of structure and functions of nature.***"

Thus, our goal in ecology is to understand the interrelations of organisms and their environments under natural conditions.

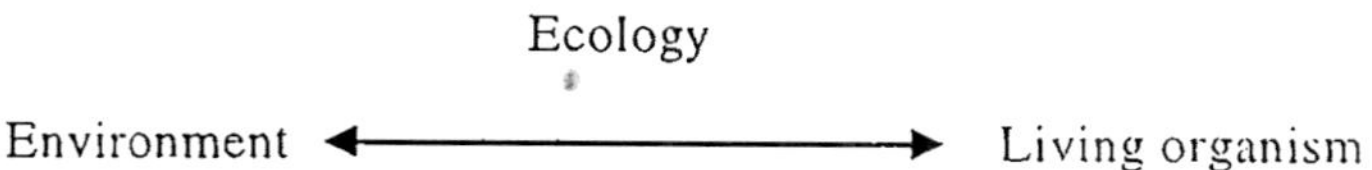

According to Clarke (t1954) the fundamental of ecology lies in the fact that to survive, animals and plants should be able to achieve three requirements from their surroundings:

(a) Regular supply of energy.

(b) A Regular supply of material.

(c) Removal of waste products.

ECOLOGY A MULTIDISPLINARY SUBJECT

Ecology is a very wide subject. It is a complex, but one of the most interesting point where many other disciplines of science and biology meet with each other and thus an interdisciplinary ground is produced . Zoology and Botany are two major subjects which gives us the knowledge of the potential biotic factors to our study of ecology. Physiology explains their functional methodology; Geology and Geography help us in knowing the earth (soil) strata along with their physical and chemical properties. The study tools and various principles are obtained from physics & biophysics; the biological activities are better understood with the help of biochemistry. On the other hand statistics and economics also contribute to ecology in the study of populations, communities and their characters. Genetics helps us to find out the inheritable and inherited characters, possible adaptive evolutionary trends among organisms and about the origin of new characters and species. The various technologies involved with the management of natural resources, survey, inventorying and control of environmental pollution, treatment of different types of waste materials have been derived from Remote sensing, Environmental Engineering and Chemical engineering. Thus it becomes very clear that ecology is such a subject- which involves the knowledge of various fully developed sciences to understand itself. In the most simplest terms ecology is the science and subject of understanding to live in perfect sustainable harmony with nature.

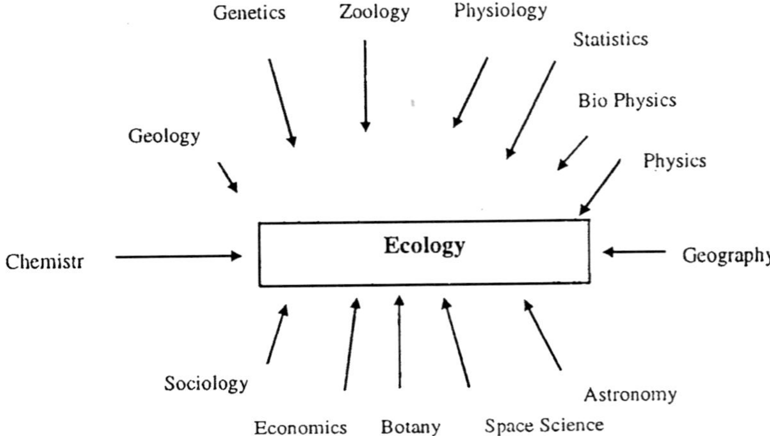

Fig. Hypothetical representation of contributing disciplines to ecology

BRANCHES OF ECOLOGY

It was only after first world war that man became much conscious about his surroundings and this was mainly on account of the fact that his ambient natural resources were either heavily polluted or became scarce. On the other hand rapid industrialization and needs of a growing human population pressed him to consider the environment and man relationship with its intricacies. It is since then that the subject of ecology and ambient environment became very important to man and, the science of ecology and environment became essential component of education in higher studies. Today the discipline of ecology is studied in various forms, with an effort to explore the precise interrelationships, mechanism of interactions between organism and its environment and their short and long term impacts on the ambient to over all global conditions and components of natural biosphere.

Important branches of ecology are listed as the following:-

1. ***Habitat Ecology:*** It deals about the nature and habitat, its characteristics and its influence on its biota (living organisms) along with various abiotic factors.

2. ***Taxonomic Ecology:*** It is another way of ecological studies, which can be further divided into branches like animal ecology and plant ecology. It can still further be subdivided into branches like invertebrate ecology, vertebrate ecology, and protozoan ecology, or mammalian ecology etc.

3. ***Population Ecology:*** This deals regarding the mutual relationships of a population in reference to other biotic components.

4. ***Community Ecology:*** It embraces a complex study of various populations co-habitating, inter-acting, co-acting and affecting each other, in a given area.

5. ***Eco-system Ecology:*** It includes all the living organisms of an area, as well as various abiotic factors both acting and reacting as well as co-acting in various ways to form the most complex grade of organization.

6. ***Palaeo- Ecology:*** It deals with the organisms of very ancient past by way of geological and geographical means and thus helps us in tracing the evolutionary events of our earth, which would further help us in understanding the future trends of our environmental components.

7. ***Conservation Ecology:*** This is the way of studying our natural resources, their proper utilization, management and maintenance. For example our forests. water resources, minerals etc.

8. ***Production Ecology:*** This branch of ecology deals about anticipated, gross and net gains of productions from various ecological resources like pisciculture, agriculture, horticulture, seri-culture, silvi-culture etc.

9. ***Human Ecology:*** This branch deals with the man and his environment.

10. ***Gene Ecology:*** It is the study of genetic changes and characteristics in the light of environmental factors.

11. ***Space Ecology:*** This is the most recent branch of ecology which deals about the study of other planetary environment, the space travels, the increasing number

of satellites in the space, possibilities of life on other planets and like that, besides the impact of space travels and events on human life.

12. ***Radiation Ecology:*** It is the study of various types of cosmic and planetary radiations, including the radio-active industries and activities of the human beings in reference to other biotic & abiotic components.

13. ***Ecological Energetics:*** This is the study of energy production, transformation, and its flow at various levels, gross and net gains from various sources viz. biological and non-biological, renewable or non-renewable.

14. ***Cyto-Ecology:*** This deals with the cellular details, characteristics and adaptations etc. in relation to its environment and its needs.

APPROACHES TO THE STUDY OF ECOLOGY AND ENVIRONMENTAL SCIENCE

To study and understand a subject of such a vast magnitude and dimensions like ecology various approaches have been put forward. The aim is to make the studies accurate, easily accessible, acceptable and useful, for the final management of our eco-system resources and its sustainability. Some of these approaches have been mentioned below.

Habitat Approach

In this method a habitat is selected and studies are made of its various components and their activities. But this becomes very complex as a habitat is a very broad term and has further sub-divisions of its own. Hence this approaches is now preferred at a very limited scale.

Taxonomic Approach

This has been other way of ecological studies which was based on the systematics of various plant and animal groups, with gradual rise of complexity in the form of a biota. This approach has its own branches like animal, plant and their further sub-divisions.

Grade of Organization Approach

Modern ecologists have arrived at a common agreement that the best way to study ecology is based on the various grades of ecological system. The ecologists have divided these various grades in the following major heads:

(a) ***Aut-ecology:*** It deals with the study of the individual organism and or individual species and its interrelationship with its ambient environment in its totality.

(b) ***Syn-ecology:*** This deals with the groups of organisms which are associated together as a unit. It can be further sub-divided according to the level of organization of the group to be studied, as following:

i. Population : A group of individuals belonging to single species or to several species, which are closely associated with each other, when studied in- mass, forms the basis of population approach.

ii. Community : It is the next higher grade of organization, where two different populations of animals or \ and plants are studied collectively. Here one tries to find out that how these two major bio-components influence each other.

iii. Ecosystem : When the living organisms and non-living physical and chemical factors of the environment, are studied collectively as an ecological unit with all their complexities, then it is called as the eco-system approach. This shows that an ecosystem is rather a most complex and highly integrated grade of ecological organization which includes all the possible factors within it, where we study a whole system or some times part of a whole system.

Productivity Approach : Under this head the resultant gains i.e. yield of an eco-system are studied, which are obtained by the energy synthesis and its transfer as well as transformation, from abiotic environment to the biotic organisms in the form of crops, fruits or flesh.

THE ENVIRONEMENT

We know that every living organism has its own immediate surrounding. This immediate surrounding or ambience consists of all sorts of non living and living components, ecologically speaking these are abiotic (i.e. without life) and biotic (i.e. with life) components. The life activities of all living organisms depend upon and are influenced with the ambient components of nature.

It is this ambience of an organism which has a very intimate, essential and mutually interacting relationship among all its abiotic and biotic constituents. This gives rise to a unique and a specific area of habitation to the organisms and is called as its environment. It can be said that the world of ecology is a market place, where all the abiotic and biotic components are available, while the world of environment is like a house, where all ecological components are there in a greater intimate and orderly manner interacting with each other for the sustenance of each other.

Our general environment is a small part of a vast ecological complex, having almost all the component of a life support system to interact and sustain itself, with all its ambient living and non-living constituents. In simple terms an environment is a small area of a greater ecological system.

The environment is not merely a source of vital requirements of any organism, but also has multi directional influences on the behavior, habit, morphology, physiology, development, reproductive behavior and genetics etc. of the organism in general. Beyond to all these, the environment demands certain adaptations and modifications from the animal towards its provisions and influences. These adaptations sharpen the living instinct and adjusting capacities of the animals with regard to the changes of the environment.

Definition of Environment: The environment can thus be defined as ***"The complex around an organism, consisting of several biotic and abiotic factors, which influence, interact, demand, on or from the organism and sustains it, through various ways of energy transfer and movement."***

According to the great ecologist Clarke (1954) environment must be satisfactory if it is able to provide the following to an organism:

i. Minimum requirements for life, and

ii. It must not exert any such influence, which may be incompatible with life.

Our ecological environment has two major components

A. ***Abiotic Components :*** This consists of space available to various organisms, their interactions and a variety of physical and chemical factors, present and distributed through air, soil and water.

B. ***The Biotic Components :*** This includes all the living organisms, flora and fauna (plants and the animals), of the living world. The detailed account of different ecological components are discussed in the preceding chapters.

SCOPE AND IMPORTANCE OF ECOLOGY

Ecology is the science of all the relations of the organisms to all their environment (Taylor, 1936). The scope of ecology becomes very diverse because of the diverse animal and plant inhabitants. The main objective of ecology, however, is to delineate the general principles under which the natural communities and their different components operate.

In the recent times, ecology has assumed greater importance due to its relationship with human beings . Modern ecology is concerned with the functional interdependencies between living organisms and their environment. It is very important to have the knowledge of principles and problems of ecology while attempting to evaluate any natural situation. A lack of principles of ecology will misdirect the efforts of well-intended conservationists and agriculturalists.

The ecological principles provide a background for further investigation in to the various relationships of the natural community and the sciences dealing with their environments like soil, forests, ocean and waters.

Ecology finds many practical applications in agriculture, biological surveys, pest control, game management, forestry, fishery biology and conserving biological diversity. Man being an organism with its own environment, has been given particular importance in the development of human ecology. Thus a knowledge of the ecology provides a background for understanding the human relations with its physical environment.

Environment as a subject has a global nature which belongs to all of us and has a closely and intricately knitted network of components and functions. The study of environment is of utmost importance as it deals with day to day problems like fresh air, safe water, healthy food and good living conditions to global concerns like global warming, ozone depletion, loss of biodiversity and energy crises throughout the world.

NEED FOR PUBLIC AWARENESS

It is very important for all of us to have knowledge and awareness regarding the state of environment around us. It becomes most important at a moment when the world is facing several environmental problems threatening to the survival of the most intelligent creature on the earth that is the human being itself. It has become imperative for the nations to educate their people so that the nature of problems being faced by them are understood and future is made secure. In this regard the use of media in creating awareness is of utmost importance. Various documentaries, advertisements, feature films and serials can be produced and telecasted on the radio, television and other means of broadcasting. Similarly newspapers and magazines should have matters related with environmental awareness.

Sustainable development, which is the major mantra of today, can not be achieved without making people in general, aware about their environment and the problems which the present society is facing because of its misuse. The masses are to be educated about the facts that degradation of our environment is actually harming to us only. At national level, the Supreme Court has recently made it mandatory for all the Universities and Colleges to teach this subject at Graduate level. At international platforms, it is being discussed throughout the world and the various issues related with global environmental concern are being taken up by the United Nations at large scale.

CHAPTER-II

NATURAL RESOURCES

A resource is anything we get from the living and or non living environment to meet our needs. Natural resources are produced by earth's natural processes. They include the air, water and land of thé planet earth, nutrients and minerals in the soil and wild and domesticated plants and animals. The natural resources have also been defined in terms of energy which is necessary for the functioning of a biological community and ecosystem. The five basic ecological variables such as (i) matter, (ii) energy, (iii) space, (iv) time and (v) diversity, are some times combined as natural resources.

RENEWABLE AND NON-RENEWABLE RESOURCES

Odum (1971) has divided the natural resources in to two groups.

1. Renewable

Renewable resources are those, which are living or biotic and are able to reproduce or replace themselves. The renewable resources can sometimes be multiplied by different methods like propagation, reproduction and grafting etc. The renewable resources are intricately linked with each other and if any single resource is influenced, the others are automatically influenced e.g., forest resources and wild life resources. Some of the resources are inexhaustible because these are being used for centauries and their supply is very huge. The most common examples include the wind energy, solar energy, tidal energy and energy derived from falling water etc. The renewable resources, thus can be modified and may be depleted, sustained or increased by human actions.

2. Non-Renewable

These are the non-living resources which once consumed can not be replaced or reproduced by the human beings. These resources are available in the nature but in limited amounts. The non-renewable resources do not reproduce. The examples of non-renewable resources include minerals and metals (gold, iron, copper, lead silver etc.) and fossil fuels (coal, petroleum, oil and natural gas). The fossil fuels account for 90% of world's production of commercial energy while hydroelectric and nuclear power account only 10%.

The Non-renewable resources sometimes are further classified in to two categories viz. Recyclable, which are mainly non-energy mineral resources which occur in the earth's crust can be collected after they are used and can be re-cycled. Examples of such resources include ores of certain metals like copper, aluminum and mercury etc. The other category of non-renewable resources is non-recyclable which includes resources which once used are exhausted for ever and can not be recycled or reused in any way. Fossils fuels are the best example of this category of resources.

CONSERVATION OF NATURAL RESOURCES

Conservation may be defined as ***the most efficient and most beneficial utilization of natural resources***. It may also be defined as the ***rational use of environment to provide a high quality of living for the mankind.*** Thus the most important aspect of conservation would be the maintenance of the widest possible diversity on the earth.

Conservation is often thought of as a twentieth century phenomenon, something we have only needed within the recent past. However, the industrial development has certainly affected to our natural resources. The rapid growth of industrial economy has increased the standard of living of people. But at the same time as the world population is increasing, per capita resource use has also increased, we all are using an increasing amount of the earth's resources to the extent that there is now global competition for them.

Aims of conservation: The conservation has following two folded aims:

(i) The conservation ensures the preservation of quality environment that considers aesthetic, recreational as well as product needs.

(ii) It ensures a continuous yield of useful plants, animals and materials by establishing a balanced cycle of harvest and renewal.

The following objectives may be specified with the living resources conservation:

(i) To maintain the essential ecological processes and the life support systems.

(ii) To preserve the biological diversity. It includes two related concepts-the genetic diversity and the ecological diversity.

(iii) To ensure that any utilization of species and ecosystem is sustainable. ***Sustainable utilization means planned utilization so that a continuous yield of the useful plants, animals and materials may be obtained.***

There are two main categories of conservation.

1. In-situ conservation

This is the conservation of living resources within natural ecosystem in which they occur. This is an ideal system for genetic resources conservation. It is achieved through a system of protected areas of different categories, managed with different objectives to bring benefit to the society. National Parks, Sanctuaries, Biosphere Reserve, Natural Resources, Monuments, Landscapes etc. belong to this type of conservation.

2. Ex-situ conservation

This is the conservation of biotic resources outside their natural habitat by perpetuating sample populations in Genetic Resources Centers, Zoos, Botanical Gardens, Culture collection etc. or in the form of gene pools and genetic storage.

In the year 1945 an international body known as International Union for Conservation of Natural Resources (IUCN) was established for the global conservation of resources. In the recent decades conservation has gained a good momentum and efforts are being made globally to conserve the natural resources left on the earth because the natural resources are fixed and limited. The existence of resources depends, basically, on its value to human beings. Therefore, to create resources, man employs technical and organizational skills along the path way of opportunity. Keeping in mind the great value of natural resources, it is imperative on our part to conserve natural resources viz. Forest Resources, Wild Life Resources, Water Resources, Mineral resources, Energy Resources, Soil and Land Resources.

ROLE OF INDIVIDUALS IN CONSERVATION OF NATURAL RESOURCES

The first great fact about the conservation is that it stands for development and the natural resources that are available on this earth are used for the benefit of people who inhabit this earth. Conservation is a practice which embraces preservation, maintenance, sustainable utilization, restoration and enhancement of the natural resources. All the natural resources including soil, water, food, forests, energy and food play a vital role in the development of a country. Their use and exploitation has, however, resulted their depletion to an extent where their conservation has become a must. Conservation is now being used in a broader sense of conserving the whole earth itself so that its self renewable capacity is protected. The conservation of fossil fuels and other non-renewable resources is now greatly demanded.

All the natural resources are conserved for their biological, commercial, medicinal, aesthetic and recreational values. The large areas under tropical forests are conserved for their role in preserving the global biodiversity and the role they play in maintaining the ecological processes, the wild life resources are preserved to maintain a genetic diversity and the different uses it is put into, different water resources like fresh water stream, is protected for natural beauty.

The processes of reuse and recycle of many of the resources helps in reduction of waste and resource consumption and also helps in the conservation of energy needed to produce the new products of consumer use. Similarly a large number of conservation methods may be used to conserve and protect the natural resources and a large number of efforts are right now under way at national and international levels in this direction. However, the individual efforts can certainly go a long way in this effort as the nvironment belongs to all of us and each one of us has a responsibility to contribute towards its conservation. Individuas can play a vital role in the conservation of different resources in the following manners.

1. To conserve the energy we must follow the following:

   (i) Build houses with provision for ample amount of sun light which will not only provide more light to the house but also keep it warmer.

   (ii) Turn off lights, fans and other appliances when these are not in use.

   (iii) By the most efficient homes, lights, cars, and appliances and evaluate them only in terms of life time cost.

   (iv) Obtain as much heat and cooling as possible from natural resources specially from sun, wind and trees.

2. To protect the water one can follow the following

   (i) Don't keep water running while brushing teeth, shaving or washing.

   (ii) Instal water saving shower heads and flow restrictors on all faucets

   (iii) Get all the water leaks repaired in toilets and pipes.

   (iv) Water saving toilets should be installed which use lesser amount of water per flush

   (v) Try to wash full loads, use the short cycle and fill the machine to the lowest possible water level.

   (vi) Sweep walks and driveways instead of hosing them off.

   (vii) Use drip irrigation system and mulch on the home garden to improve irrigation efficiency and reduce evaporation.

3. To conserve wildlife and forests:

   (i) Do not harm or remove while visiting to a national park or sanctuary and our moto should be leave it better than we found it.

(ii) Support efforts to bring more areas under the protected areas.

(iii) Restore damaged forests, grasslands, croplands wetlands and strip-mined lands.

(iv) Plant trees in rural and urban areas.

(v) Reduce soil erosion

(vi) Help establish and maintain wildlife refuses for endangered plants and animals.

4. To protect the soil:

(i) Do not up root the soil as far as possible and plant the fast growing native plants in the disturbed areas.

(ii) Grow grass in the open areas to bind the soil and prevent the soil erosion.

(iii) Make compost from the kitchen waste and use it for kitchen garden or flower pots.

(iv) Use sprinkling irrigation as it does not wash off the soil.

(v) Use mixed cropping so that some specific soil nutrients do not get depleted.

(vi) Do not over irrigate agricultural fields without proper drainage to prevent water logging and salinisation.

(vii) Promote sustainable agriculture

(viii) Reduce the use of pesticides

(ix) As far as possible use organic fertilizers

(x) Grow local and seasonal vegetables.

EQUITABLE USE OF RESOURCES FOR SUSTAINABLE LIFE STYLE

The economic indicators of growth are flawed because they do not distinguish between resource uses that sustain progress and those which undermine it. The over use of resources by a few at the expense of others, whether of water, forests, land or minerals etc. are reflected as increase in Gross National Production even though it increases the misery of many who are left behind. In India inspite of the increases in GNP, in the 1960s, there were 17000 villages in Uttar Pradesh plagued with water shortages. In 1985 they had increased to 70,000 villages. In Gujarat in 1979 there were 3840 villages with water shortages, in 1986 this number increased upto 12, 250 villages.

There is a big divide in the world as north and south, the more developed countries and less developed countries, the have's and the have nots. The gap between the two is mainly because of population and resources. The more developed countries have only 22% of world's population, but they are using 88% of our natural resources, more than 73% of energy and command 86% of its income. The low developed countries on the other hand have very low or moderate industrial growth, possess 78% of world's population and use only 12% of natural resources, and 27% of energy. With the rapid increase in the population of developing or low developed countries, the gap between the rich and poor is increasing with passage of time.

In India more than 75% of our population lives in rural areas with more than half of them having no lend, employment and are oppressed, the development bias in terms of investment, education and training and infrastructure development, supports the urbanized population. Even after the 60 years of it's independence, a large part of rural India is without drinking water, schools and health facilities, roads, electricity or other energy resources are also lacking.

The UNESCO declaration that "Since wars begin in the minds of men it is in the minds of men that defenses of peace are to be constructed" is equally relevant about poverty, environment and sustainable development issues. It is in the mind of men that issues of justice, human rights, sharing of resources will have to be fought to ensure a better world.

CHAPTER-III

FOREST RESOURCES

A plant community dominated by trees and with other vegetation usually with a closed canopy is called forest. There are no reliable statistics of forest cover in India, however, at the time of independence, the recorded forest area of the country was reported to be about 40 million ha as per the record of Central Statistical Organization. The government owned forest was about 26 million ha and community and privately owned forests were about 14 million ha. The area increased to 68.02 million ha in 1950-51 with the addition of ex-princely and ex-proprietary forests. The area further increased in the early eighties to 75.18 million ha. Presently the recorded forest area is 76.52 million ha. In India 16 different forest types have been described by Champion and Seth (1968). According to them a forest type is a unit of structure sufficiently pronounced to permit its differentiation from other sub units.

Among 16 different forest types of the country the most common is tropical dry deciduous (38.1%), followed by tropical moist deciduous forest (30.9%). The following table gives the status of different major forest types in the country:

Forest Types	Percentage of Total
1	2
Dry deciduous	38.1%
Moist deciduous	30.9%
Tropical thorn	6.9%
Tropical dry ever green :	0.1%

1	2
Pure coniferous (high mountainous area	6.3%
Sal forest	16%
Teak forest	13%
Broad leaved (excluding sal & teak)	5.8%
Bamboo's (including in plantations) :	8.8%

Out of the total forest area, nearly 96% is owned by the government, 2.6% by the corporate bodies and rest are in private ownership. The exponential growth in the population has resulted a decrease in the forest cover, though traditionally Indians have a deep involvement with forest and forestry. The National Forest Policy 1952 enunciated that 1/3 of the geographic areas of the country should be under the forest cover. However, there had been continuous deforestation in the country for various reasons and it is estimated that 4.238 m ha forest land was officially diverted for non forest purposes between 1951-52 and 1979-80.

USE AND OVER EXPLOITATION OF FOREST RESOURCES

A forest is the most effective eco-system that supports many kinds of organisms and absorbs the energy of the sun to build up organic materials. Forests are the areas set aside for the production of timber and other forest produce. Forests play an important role in maintaining environmental stability and in supplying the essential requirements of almost all the living organisms including the man. Forest provide us a large number of products which have a commercial and ecological values.

The chief economic products of forests include timber, fire wood, pulp, food items gums resins. The various products like oils, rubber, fodder, medicines, drugs, lac, canes, bamboo are obtained from forests only. Thus the forest products have a direct commercial use without which the human societies can not sustain.

The ecological importance of forests in terms of purifying the air, mainating the water table and thick humus layer, reducing the global warming by absorbing the carbon -di-oxide gas, binding the soil layers and controlling the soil erosion are the other important uses of forests to which we can not fix a price tag. Similarly, forests provide a natural habitat to millions of wild animals and plants. Forest also have a recreational and aesthetic value for human beings.

Initially, the forests were seen as an almost inexhaustible resource. But due to increasing human and livestock population, the forest resources of the country have been under mounting pressure. Increased urbanization, industrialization and mining has caused indiscriminate felling of trees and denudation of forests. The practice of logging down the trees for human use and over grazing of forest areas have been growing with the passing of centuries. Colonists used logs for firewood and making huts. Humans used the logs and other forest produce for various purposes.

DEFORESTATION

A most recent report says that our earth is being denuded of 5 hectares of forests every second, which means that 157.5 crore hectares of forest being obliterated every year. According to Brewbaker (1984), the total forest area of the world in 1900 was nearly 7000 million hectare, by 1975 it was reduced to 2890 m ha. The major reduction is in tropics and sub tropics (40.2%), while it is only 0.6% in temperate areas. In the tropics and subtropics, maximum reduction would be in Asia and the Pacific. This is mainly due to the explosion in the human population. The problem of forest degradation is highly alarming in the Asian countries. Important causes of forest degradation are as follows:

(1) Forest fires

Forest fire is one of the most important ecological factors in the terrestrial environment. It has affected a substantial portion of the forests of the world. The fires may be ***intentional***, which are set intentionally. Sometimes miscreants set fire to the forest with an intention to damage the forest wealth. In Himalayan area, villagers set fire to the forest floor so as to induce a good growth of grass in the following season. This is an effect of over-population of cattle in this area. Controlled fires are also set by the Forest Department in some cases, but such fires hardly produce any harmful effect on the environment. Tourists and poachers may cause ***un-intentional fires***, who may leave burning cigarette butts, matches, campfires etc. in the forest. Unintentional fires are also caused by lightning.

Fires kill the vegetation of all types or injure them severely, ground fires burn the undergrowth of the forest and some of the underground parts of the plants. Direct effects of fires include the burning of biomass, release of carbon dioxide and nitrogen gases and ash in the atmosphere and deposit of nutrients.

(2) Timber extraction

Tree felling was started for the domestic use and for the manufacture of agricultural implements and for the developmental activities such as for the supply of railway sleepers.

Wood smugglers in different parts of the country do the illegal felling. Villagers also remove illegally some of the forest produce in the form of fuel-wood or grass. Most villagers enjoy certain rights for collection of dry stems, barks, leaves, etc. from the neighboring forest for their energy needs. However, dry twigs, leaves etc. from an average forest is seldom enough to supply the fuel wood needs of a village having about 40-50 families. The villagers are forced in such circumstances to extract more fuel wood from the forest, than is allowed under the rules.

Contractors sometimes cut down more trees than that is allowed. This practice was common during 1950s and 1960s. People used to enter the forests illegally and remove one or two trees. Some nomadic tribes in the Western Himalayas also do illegal felling of trees. They live in the forests and cut down the forest illegally. They earn their livelihood by selling milk to the nearby towns and village. The cumulative effect of these activities resulted in the forest degradation. However, with the abolition of the contractor system in most areas and introduction of very strict laws against illicit felling, this activity has been checked. But poachers and timber-wood smugglers in different parts of our country, which needs immediate attention and control through strict legislation and watch, are still doing the illegal felling. India alone is losing more than 1.5 million hectare of forest cover each year and 22 million hectare of forests was destroyed during the three decades of 1950 to 1980 (Tewari, 1982).

(3) Overgrazing

Overgrazing by cattle is one of the greatest factors in the degradation of pasturelands particularly in the Himalayan areas. Over the years, with the increase in cattle population, there has been a rapid degradation of the forest environment. In J & K, after 1970 the cattle population nearly doubled. 20% of the total cattle population of the country is in the Himalayan region, which has about 18 percent of the forest area of the country.

The cattle population is maintained for the following reasons:

(i) Milk and Agricultural Uses: Cows, goats and buffaloes are reared for their milk in different villages. Some villages also depend upon Gujars or nomadic/semi nomadic tribes for their milk requirement. Bullocks are used for ploughing the field.

(ii) Manure: Animal dung has a prominent place in hill agriculture where chemical manures are not in much use. Each farmer likes to keep about 10 animals per hectare of cultivation in the hilly areas. Many farmers also keep one or two stall-fed buffaloes, but stall-feeding of cows is rare. Usually, animals kept exclusively for this reason are allowed to graze on the pastures during the day and they are kept in enclosures during the night.

(iii) Wool: Some nomadic and semi-nomadic tribes in the hills and foot-hill zones rear goats, sheep and yaks for their hair, which is used to make woollen garments, blankets, carpets etc.

Ecological effects of over-grazing are as follows:

1. Over-gazing causes repeated grazing of grasses and forbs in the pasturelands, which affects their regeneration and growth.

2. Most of the species by repeated grazing develop into prostrate plants, which affects their normal growth and tolerance.

3. The carrying capacity of the pastures is not followed in overgrazing practice. When the number of cattle

exceeds the carrying capacity of a pasture, these develop into degraded lands.

4. The plant species susceptible to repeated grazing are eliminated from the pasture, only the resistant species remain there. This affects the normal community status of such over grazed pastures.

5. Due to repeated movement of cattle in the same pasture, gullies are formed in their tracks which accelerate the soil erosion and land degradation

6. Cattle hooves also damage Young regeneration, seedlings and saplings of trees and shrubs.

(4) Mining and dam construction

Mining is a biotic factor in which man is exploiting the resources of nature, especially the minerals. This activity is done for various purposes such as mineral extraction, road construction and major hydroelectric projects. These activities have reduced the mineral resource itself and at the same time have caused considerable adverse changes in the ecosystem. The extraction of minerals and materials requires removal of vegetal cover; soil mantle and excavating overlying rock masses (overburden), which commonly exceed the volume of material extracted. Some of the important minerals, which are extracted by mining, include limestone, dolomite, phosphorite, gypsum, graphite and magnesite.

Surface or open cast mining causes a considerable loss to vegetation and soil cover particularly in the Himalayan areas. Mining affects the distribution of plant species, reduces the vegetation cover and habitat, and causes the loss of soil cover and increases the soil erosion. This activity sometimes results into the instability of slopes in the hilly areas, landslides and silting-up of rivers. The abandoned mines are invaded by exotic species such as Lantana camara, Euphorbia royleana, Eupatorium species, etc. These lands ultimately get converted into wastelands or highly degraded lands. Mining activity also affects the survival of native species such as orchids and some medicinal plants.

The construction of large dams are also responsible for destruction and degradation of large areas which are under forests.

In India we have about 1556 dams indifferent states and maximum being in Maharashtra. Dam building requires a large scale devastation of forests and conversion of large forest land in to a dam site. Large dams with a height of more than 100 meter are generally constructed near forest areas. Tehri Dam is the most recent example of high dam where a large number of families were to be rehabilitated and forest and agriculture land was submerged under more than 100m of water. Tehri dam on river Bhagirathi is the highest dam in the Asia while Bhakra dam on river Satluj is the largest in terms of capacity.

CASE STUDIES

The Tehri Dam Project

Investigations on Tehri, which was first to be locatedat Dobra, 12 Km. upstream to Tehri was completed by mid sixties. But later the site was shifted below the confluence of Bhagirathi and Bhilangana. The installed capacity of the dam was to be 600 MW (1972), irrigation to 2.7 lakh hectare and cost was 1977 crorore rupees. But the project was revised and it was to generate 2400 MW electricity and its costing 1989 was Rupees 3,000 crores, which has now been estemitated to be more than 7000 crores of rupees. The lake created by the construction of dam submerge nearly 100 villeges, including Tehria historical capital. As many as 85,600 families will be relocated. The project intends to supply 500 cusecs of water to New Delhi.Tehri Dam project provoked controversy on three issues. The completed dam will displace many people and submerge several towns, among them the Town of Tehri; the region is vulnerable to earthquakes and the dam may be structurally incapable of withstanding them or may perhaps cause them; and the possible failure of the dam could kill hundreds of thousands of people and destroy down stream towns.

The project was challenged in the Supreme Court of India by Tehri Bandh Virodhi Sangrash Samiti (TBVSS) in 1985 . The Indian National Trust for Arts and Cultural Heritage (INTACH) sponsored an independent assessment of the economic feasibility of the dam. It was found that the benefit to cost ratio of the Tehri dam, after calculating social and environmental costs and benefits, works out to be 0.56:1, well below the 1.5:1 ratio adopted by the Plannining Commission to sanction such projects.

The Supreme Court of India dismissed the petition in 1990. Although the Environmental Apprisal Committee had unanimously concluded that the Tehri project should not be approved.

(5) Shifting cultivation (Jhum)

Shifting cultivation is a practice, which prevails throughout the world, particularly in the hills inhabited by tribals. In this practice, the cultivator does not use a particular piece of land year after year. A patch of land is selected, all the trees, herbs and shrubs are cut down at or near the ground level there, which are left to dry in the sun and then are set on fire. The cleared and burnt areas are then taken up for cultivation. Seeds are sown in the ground. This cultivation practice is carried out by human labor without ploughing the land or using animal labor.

Shifting cultivation is perhaps the most primitive form of cultivation. Over 2.6 million people belonging to tribes who live in the interior hilly area practice shifting cultivation in India. About 1.35 million acres of land is under this form of cultivation in India in Assam, Arunachal Pradesh, Andhra Pradesh, Meghalaya. Mizoram, Nagaland, Manipur, Tripura, Bihar, Orissa, Madhya Pradesh, Karnataka and Kerala. It is called Lo or Jhum in Mizoram and Assam and by other names in other parts of India.

Main crops raised include grain crops and also cash crops. Among the food grains, coarse varieties of rice is the principal crop, others are maize, millet, jowar etc. Vegetables include pumpkins, cucumber, melons etc. Cotton is also raised as a cash crop. One important feature of such areas is that the land under shifting cultivation does not get any replenishment of nutrients in the form of manure. Thus, yield from these areas starts diminishing from the second year onwards. The cultivators then move to a fresh patch of land. The cultivator usually does not return to the same land again.

The major forest types in which shifting cultivation is practiced in the eastern Himalayas are tropical wet evergreen forest, tropical semi-evergreen forest, tropical moist deciduous forest, subtropical broad-leaved forest, subtropical pine forest, montane wet temperate forest and Himalayan moist temperate forest.

(6) Tourism

Tourists are of different types based on their purpose such as pilgrims who visit a number of shrines and religious places, tourists who visit seasonal resorts, and the mountaineers and adventure sports-persons who visit the Himalayan and other hilly areas. The major environmental problems caused by tourism are as follows:

1. Tourism activity leads a pressure on the forests. Due to large number of tourists, the grasslands and forests are destructed by horse riding and by fires, which are caused accidentally. The horse owner makes a pressure on the forests for the fodder requirements of their horses during the season of tourism or pilgrimage.

2. Health and sanitation problems are caused by tourism during the peak influx of tourists. The maintenance of eatables and drinking commodities becomes a tidy job and sometimes make these commodities unhygienic.

3. This activity is also causing the pollution problems in the tourist areas. The mountaineers leave behind much garbage in the high altitude areas, which is a matter of concern these days.

(7) Natural causes of forest and land degradation

Besides the biotic causes, natural processes such as earthquakes, glaciers, mountain streams etc. too are creating environmental problems related to forest and land degradation. These processes are as follows:

Earthquakes: Earthquakes are also responsible for land degradation. This activity is highly destructive in the Himalayan and other hilly areas. Some of the important earthquakes, which took place in, the Himalayas and other parts of the country, are as follows:

(a) Kashmir earthquake (May, 1885)

(b) Assam earthquake (June, 1897)

(c) Assam earthquake (August, 1950)

(d) Kinnaur earthquake(January, 1975)

(e) Dharchula earthquake (1980)

(f) Uttarkashi earthquake (October, 1991)

(g) Chamoli land slides and earthquakes

(h) Latur earthquake

(i) Bhuj, Kutch earthquake in Gujrat, 2001.

(j) Kashmir earthquake (2005)

Earthquakes cause severe damage to land and forest in addition to the human life and animals.

Glaciers: A glacier may be defined as naturally moving body of large dimension, made up of crystalline ice formed on the earth's surface as a result of accumulation of snow. Glaciers are a major source of erosion in the snow bound areas. They form U-shaped valleys, hanging valleys, crevices, cirques, glacial scars etc. by erosion and lateral, medial, englacial or ground moraines, drumlins etc. by transportation and deposition.

EFFECTS OF DEFORESTATION ON FORESTS AND TRIBAL PEOPLE

The different effects of deforestation include:

1. Clearing and burning of vegetation reduces the forest cover to the loss of timber worth lakhs of rupees and and loss of other forest produce.

2. The natural habitats of wild flora and fauna are depleted as a result the loss of diversity takes place.

3. Damage to the vegetation cover leads to soil erosion, silting up of rivers, reservoirs, depletion of soil nutrients, floods, etc.

4. The Jhum cultivators accidentally may set fire to the adjoining forest, which may affect the vegetation and its regeneration.

5. The abandoned cultivated areas remain fallow land for 10-15 years, which means a great loss as productive land is kept useless.

6. The rare or sparsely distributed plant species are destructed by the different practices of deforestation.

7. The tribal people who are dependent on forest and forest products are the worst sufferers as the sources of their livelihood are destroyed for ever.

8. Large areas of forest are used as grazing lands by the communities living close to forest areas and once the forest land is converted in to different land use pattern, the pasture or grazing land does not remain available as a result the economy of the community is adversely affected.

FOREST CONSERVATION

In India, the first forest policy was declared in 1894. This was revised in 1952. This had recommended a national target of 100m ha of forest and suggested that one third of the geographical area of the country should be under forest. The government of India enacted the Forest (Conservation) Act, 1980 with a view to check indiscriminate-devastation and diversion of forest land to non-forest land or non-forest purposes. The forest (conservation) act 1980 was amended in 1988 to incorporate stricter penalties against the violator of the act.

In the year 1987 the first State of Forest Report was published giving actual forest cover of the country. So far the forest cover assessment of the country has been made on six occasions using satellite data. The govt. has already setup a National Forest Fund. In 1990, guidelines have been issued to State Governments for involving village communities and voluntary agencies for regeneration of degraded forest lands on sharing basis. Social forestry, urban forestry, restrained felling, selective and block cutting and afforestation are some of the methods used for further conservation and management of forests.

1. Social forestry

Social forestry is simply defined as the forestry of the people, by the people and for the people. Social forestry means the management and protection of forests and afforestation of barren lands, with a purpose of helping in the environment, social and rural development, as against the traditional practice of securing revenue. The National Commission on Agriculture (1952) has classified social forestry into three broad clàsses viz. urban forestry, rural forestry and farm forestry.

Types of Social Forestry: The types of social forestry are as follows:

(a) Farm Forestry

This type of social forestry includes raising of wind breaks, Shelter belts, farm wood lots, raising trees in village common landsetc. Windbreaks and shelterbelts are planted in the dry land agricultural areas. The higher speed of wind causes rapid evaporation of water. To control the effect of high velocity winds, the plantations are raised in the form of rows, which are called shelterbelts. These belts consist of rows of trees and shrubs, planted along the field borders at right angle to the prevailing high velocity winds. The belts are 3 to 5 rows of trees and with a conical cross section, that is, towards the windward side shrubs and bushes are planted close to each other, the second row will have medium sized trees and the third row tall trees, and then in the declining order. The common examples of plants in such belts are Agave, Falsa etc. in the first row, Sesbania, kubabul, etc. in the second row and third row with Eucalyptus, Casurarina, Acacias, etc. The shelterbelts reduce the wind velocity over the crops and also reduce the evaporation losses. The increase in dew formation in these belts is suitable for the good growth of some crops like sorghum. The belts also control the soil loss from the crop and deposition into the crop area. The planted trees also yield fruit, fodder, fiber, fuel wood and small timber.

The windbreaks are the rows of trees planted along field margins in the perpendicular direction to the most high velocity winds. Roughly seven percent of the total area should be planted for effective protection. The windbreaks are raised by planting two close rows of fast growing deciduous and one parallel row of slow growing and longer living evergreen species (Tamarind). Other trees suitable for-wind breaks are Casuarina, Melia, Tamarix, Eugenia, Eucalyptus, Artocarpus, Grevillea, etc. Anacardium, Acacia, Casuarina, Agave, etc. are suitable for coastal areas. Shrubs and grasses are also used in these breaks.

Wood lots and afforestations are raised on lands unsuitable for cultivation. Every village must have at least 10 percent of its cultivated area relegated for this purpose. These wood lots will be the property of the whole village and the benefits should be utilised for the betterment of the village.

(b) Extension Forestry

This also includes agro-forestry practice. In extension forestry raising tree crops on canal banks, railway lines and road sides, under high tension electric lines, irrigated plantations of forest trees, afforestation of foreshore areas of tanks and reservoirs, reclaiming lands unsuitable for agriculture with tree crops (agro-forestry) are done.

Aforestation of canal banks is an important practice in social forestry because with the development of irrigation and power projects, these banks are also increasing and are creating new sites for plantations. These are poor in fertility and need to be protected from erosion, which starts with rills and slowly gullies are formed and these create breaches in the canal. Tree growth protects the banks and also prevents evaporation of water in the canal. Canal bank plantations also serve as shelterbelts to many fields adjoining the canals. Economic and shade providing species are preferred in the canal bank plantations. These may yield fuelwood, pulpwood, fodder, fibre and timber. Bamboo and teak if raised in suitable well-drained soils will prove of high economic value. Bamboo can be harvested every three to four years. The other species include Eugenia, Dalbergia, Mangifera, Terminalia, Tamarindus, Azadirachta, Eucalyptus, etc.

Afforestation of railway line sides and under high-tension lines is done to make these areas green, which could reduce the noise pollution, caused by trains and also the heat, making the journey comfortable. These plantations also check the erosion along these lines. The species selected for such planting should be from the local areas as they can withstand the climatic conditions prevailing there. The species planted along railway tracks include Agave, Sacharum, teak, Eugenia, Acacia,etc.

Under high tension electric lines the species, which are, short and do not interfere with the power lines should be selected to utilise such lands. Lemon grass, baib grass, Vinca rosea (a medicinal plant) and Eucalyptus citriodora (maintained as a bush) are the favourable species.

Afforestation of roadsides is done to maintain the ecological balance, to provide tree shade to the travellers, to obtain forest produce, to reduce the noise pollution, to decrease the soil erosion and for aesthetic value. The species suitable for such plantations are shisham, Terminalia, Albizzia, Delonix, Cassia, Erythrina, etc.

Afforestation of foreshore areas of irrigation Projects and ponds in dry farming areas is another type of extension forestry. In the areas of irrigation projects the top soil is removed and the rocks get exposed. Such areas should be planted by hardy species such as Eucalyptus, Acacia, neem, Albizzia, teak, mango, etc.

For uncultivable and wastelands, reclamation is done to revegetate them. In bare areas, the planting is done in pits. In eroded areas trenches are made and gully plugging is also done. Species of Cassia, Albizzia and Vitex are suitable species. The practice of agro -forestry is also done in the areas where the crop could be grown between the rows of planted trees.

(c) Urban Forestry

This includes the plantations of tees in urban areas as a forest or amenity planting to prevent and minimize pollution. Urban areas are affected by smoke, noise and dust whose adverse effect cannot be easily estimated. Sometimes roads and paths have narrow margins; therefore, tree planting should be done wherever conditions are favorable and suitable. Tall narrow crowned trees are selected for such situations. In some urban areas planting is done in parks or green belts. Species suitable for such plantings include Cassia, Delonix, Ficus, Mango, Eugenia, Jacranda, etc.

The amenity planting is done in the areas of public buildings, schools, colleges, hospitals, offices, monumental and historical gardens, crematoriums, cemetaries, industries and home gardens.

2. Restrained felling

The concept of sustained yield in forestry denotes the management of forests in such a way that a modest timber crop may be harvested indefinitely, year after year, without being depleted. In such a case the annual decrements should be counter balanced by annual increments. Maximum limit of annual felling should be fixed and the annual growth should be increased through different silvicultural practices.

3. Selective and block cutting

Block cutting method is suitable for the forest having almost even-aged trees of a fewer number of species. In this method trees are harvested not at random but in blocks. For example, from 10 acre of such forestland 1-acre block may be harvested each year for 10 successive years. If each harvested block is receded successfully, 10 age classes would be available after this period, for the blocks 1 to 10 consisting of 1 to 10 years of age respectively. After the 10 year period 1 acre of 10-year-old stem may again be harvested each year. In block harvesting method, the ecological, aesthetic and recreational uses are not maintained.

The selective cutting involves the cutting of some of the selected trees from a mixed forest or from a patch having trees of different age classes. Dense patches are selected first to mark the well grown trees for harvesting. This practice is done for the whole forest without clearing a particular portion or block. This practice is carried out on year-to-year basis.

4. Reforestation

The forests harvested by block or selective cutting methods are reforested. The bare areas are subjected to planting or reforestation. Sometimes, the seeds from the trees left after cutting become dispersed, but this may or may not be uniform in distribution for raising the new crop. Therefore, artificial seeding is required for adequate management of the forests.

5. Through Policy and Law

It is the endeavour of the Govt. of India to bring 33% of the total geographical area of the country under the forest cover. The various policies and laws made and passed by the Parliament of India are a step forward to achieve the maximum protection and conservation of forests of the country.

CHAPTER-IV

WATER RESOURCES

Water is our most abundant resource, covering about 71% of earth's surface. There is about 97% of salt water and rest 3% is fresh water. Essential to all life, water constitutes from 50 to 95% of the weight of all plants and animals. It is also essential to agriculture, transportation and countless of other human activities. Out of the 3% of the fresh water, three fourth is not available for use by plants and animals including the human beings because it lies too far under the surface of earth or locked up in the glaciers, polar ice caps atmosphere and soils. This leaves 0.5% of earth's water available as fresh water in rivers, lakes and economically recoverable underground deposits. Water, which is an essential element of life, is found in a number of spring and perennial rivers, which have their origin in the Himalayas in the north-east India and other mountain ranges in other parts of the country. The springs, though extensively distributed and provide drinking water for a fairly large number of villages, situated on high hill tops, are not, so far as is known, large enough to permit their scientific management and utilization for any large scale drinking water supply scheme. However, a limited location such as in Kashmir, large springs provide copious discharge and can be developed for drinking water supply. In any scheme of preserving and developing the Himalayan environment, the detailed studies of springs can yield fruitful result.

SURFACE AND GROUND WATER USE AND OVER EXPLOITATION

Surface water

The water coming through precipitation like rain and snow when does not percolate down in to the ground or does not return to the atmosphere by evaporation or transpiration processes of water loss remains as surface water in the forms of streams, lakes, ponds, wetlands and artificial reservoirs. It is the most important form of water resources available to human beings and put to use for irrigation, industrial, navigation and public supplies purposes. Thus it is helpful in increasing the economics of a country.

A large number of perennial river systems originate and flow down the Himalayan slopes and these provide a very potential opportunity for the development of drinking water and irrigation supply and for the generation of Hydro electric power. In implementing these developmental schemes, however, it is necessary to ensure that adverse environmental impacts do not arise. The Himalayas play an important role in replenishing the water resource of our country. This is by their role and acting as a barrier range for the monsoon winds & thus enabling precipitation of their moisture as snow or rain on the southern slops. The resulting runoff not only feeds the Himalayan drainage system but also recharges many of the aquifers, lying in the Tarai Bhabar and Indo-Gangetic plains, thus spreading the benefits over a very large area and to a large segment of our population. In providing large storage in the Himalayan Rivers System for permitting irrigation and power development, the efforts should be made to see that the ultimate benefits occurring by ways of ground water recharge and supplementation of ground water storage are not adversely affected, thereby leading to a degradation of the human environment in these places.

The Himalayan glaciers also represent a large reservoir of fresh water resources. The total number of glaciers in the Himalayas is about 15,000. Von Wissmen has estimated that the glaciers represent an ice cover of the order of 33,000 $km.^2$ representing nearly 1,200 $km.^3$ of ice. So far only estimates of this resource were available, but in recent years, the logical division of Z.S.I., located at Lucknow and Calcutta have taken up systematic and scientific studies of the Himalayan glaciers. These studies are invaluable for the management of water resources and for monitoring the environmental changes brought about by the glaciers in terms of temperature variation and pattern of wind distribution.

Ground water

The water which is received through precipitation (rain, snow and dew etc.) percolates into the ground and supplements the underground water reserves. According to an estimate about 9.86 % of the total fresh water resources are in the form of ground water. It is considered as a very pure form of water, however, in recent times due to various anthropogenic activities, this form of water has also been polluted.

Although we can not increase the earth's supply of water, we can manage what we have more effectively to reduce the impact and spread of water resource problems. There are two major approaches to water resources management: Increase the usable supply and, decrease unnecessary loss and wastage. Most water resource experts believe that any effective plan for water management should rely on a combination of these approaches.

Some rain water and water from melting snow that would otherwise be lost can be captured by dams on rivers and stored in large reservoirs behind the dams. This increases the annual supply by collecting fresh surface water during wet periods and storing it for use during dry periods. In addition dams control the flow of rivers and can reduce the danger of flooding in areas below the dam and provide a controllable supply of water for irrigation. Storage of water behind the dam also raises the water table.

Water conservation can further be achieved by making people more aware to prevent the unnecessary wastage of water. The reoccurrence of floods and drought, improper use and pollution of water and its resources needs to be checked The recycling process of waste water is promoted and at the same time the qualitative degradation of water is checked.

The different uses to which water is put into includes: drinking, cooking, washing and other domestic purposes; for irrigation of agriculture and for industrial uses; for promoting fisheries and other aquaculture practices; generation of hydroelectric power, for navigation and recreation. However, with increasing human population and rapid industrial development the demand for water has increased several folds. In India, where the Agriculture is the main occupation of more than 50% of its population, 93% of water is used for irrigation and other agriculture practices. This is incontrast of the countries like Kuwait, a water poor country, where only 4% water is used for irrigation purposes. According to a report of United Nations in 2002, throughout the world 101 bilion people do not have the access to safe drinking water and 2.4 bilion do not have adequate sanitation facilities. The magnitude of water pollution has gone up due to various anthropogenic activities viz. addition of agricultural, municipal and industrial wastes.

FLOODS

A flood is a high flow of water which over tops either the natural or the artificial banks of a river. The most unusual cause of flood is a period of heavy rain. Floods caused by storm surges or the melting of snow and ice are also caused due to the atmosphere. The colaspe of dam also lead to flooding. River floods due to heavy rainfall occur because too much water gets into a river in short span of time, as a result the water rises above the bank levels and surrounding areas are flooded. The frequency of floods is increasing and this could be due to either physical or human factors. However, the most vulnerable landscapes for flooding are low lying parts of active flood plains and river estuaries.

The countries like India and Bangladesh experience the reoccurrence of floods in many part every year. These countries receive 90% of the rainfall in a span of three to four months and prolong downpour causes the over flowing of rivers and streams resulting into floods. These floods cause a huge loss of agriculture, property and sometimes human lives. Assam, Bihar and Uttarpradesh are the worst affected states in India, where loss of property and human life takes place every year due to heavy rains and floods.

DRAUGHTS

Drought may be defined as any unusual dry period which results in a shortage of water. Thus the deficiency of rainfall is the triggering factor, but basically it is the shortage of useul water in the soil, in rivers and reservoirs which lead to acute problems. All over the world about 80 countries are lying in the arid or semiarid zones and experience the frequent spells of draught. Situation become worse in these countries because these are the countries with very high population growth thus leading to a poor land use. A variety of draught typesave been recognized, which include: meterological draughts dominated by shortfall of precipitation; hydrological draughts which mainly affect the water resources and urban water supplies; agricultural draughts are characterized by widespread regional effects in the developing countries and mainly affecting farm production and the famine draughts confined to the least developed countries which are dependent on subsistence agriculture.

Sometime the human activities result in the process of a draught. Erroneous and intensive cropping pattern and heavy exploitation of water resources to get high productivity lead to the phenomenon of draught. Draughts develop over a period of months and have prolonged existence over a period of years for major events.

CONFLICTS OVER WATER

Demand for clean water caused by surging population growth, environmental abuse and poor water management is becoming a dangerous source of friction in many parts of the world. Issues related with sharing of water particularly the river water have been largely affecting a large number of people specially the farmers and have become a great issue which is sometimes shaking the Governments.

The middle east, one of the world's perennial war zone, has traditionally been blessed with huge sources of oil, however, it has been cursed by scarcity of water, as has been rightly said that when ever a country in this area digs for water, it invariably strikes oil. The longstanding speculation among different political experts is that the world's future wars will be fought over water, not oil.

Sunita Narain, the winner of the 2005 Stockholm Water Prize says "Water wars are not inevitable. It lies in our hands- and in our minds. Water is a replenishable commodity. The question is society's relationship to live with water. The management of water is critical. Water wars and water peace is in our hands." She admits that water stress leads to tension and conflicts- as evident by a recent police shooting of farmers in Rajasthan, India. The farmers were protesting the release of water from their lands to neighbouring cities. It was a very violent agitationion.

In a paper about the Tigris and Eupharates rivers, which flow through Turkey, Syria, Iraqu, Iran, Kuwait and even northern Saudi Arabia, Prof. Olcay Unver of Ohio- based Kent State University says that despite the political volatility of the issue, shared water resources management between Turkey, Syria and Iraqu may promote international cooperation, as opposed to interstate conflict, in the coming decade. The sharing of water is also an ongoing dispute between Israel and Palestinians living in occupied territories. UN figures suggest that there are about 300 potential conflicts over water around the world arising from squabbles over river borders and the drawing of water from shared rivers and aquifers.

The Cauvery water dispute is one of the issues which has drawn the attention of different organizations. It is an issue which is yet to be solved by the Governments of Karnataka and Tamilnadu and the fighting is over 100 years old. Tamilnadu, occupying the down stream region of the stream wants water use regulated in the up-stream, whereas the Karnataka, the up-stream state refuses to do so and claims its primacy over the river as upstream user.

The Yamuna water is another bone of contention between Haryana, Uttar Pradesh, Delhi and Himamchal Pradesh. Similarly the issue of sharing Ravi-Beas waters between Punjab and Haryana is pending in the Supreme Court of India.

DAMS: BENEFITS AND PROBLEMS

The big dams have always been considered to play a vital role in the development of a country. It is more important incase of India as it has the highest number of river valley projects, where Dams have been regarded as a symbol of country's development and the local, people living close to these places develop big hopes on such projects basically for employment which will raise the living standard and quality of life. However, if we take the example of Tehri Dam, it appears that it has always been a very controversial issue. And now when the tunnels are being closed to fill the dam fresh problems are coming up.The tunnels were closed after the Union Power Ministry gave the green signal. The water level in the valley has risen to 30 metres and the closure of two more tunnels, which are at a higher level, will lead to the complete submergence of Tehri town. The closure of these tunnels has been delayed owing to unresolved issues relating to rehabilitation.

Sunderlal Bahuguna, environmentalist and founder-leader of the Chipko movement went on a fast protesting against the inadequate measures taken to rehabilitate people who would be affected by the construction of the 260.5-metre-high earth-and-rockfill dam. The construction of the dam is expected to result in the formation of a reservoir over 42 square kilometres, fully submerging Tehri town and about 20 villages and partly affecting 74 villages. For Bahuguna, the support of "anybody and everybody" was welcome as long as it met the aspirations of the displaced people of Tehri. The protest, after all, was hinged on an emotive issue that was divorced from the more realistic problems such as displacement and rehabilitation. The construction of the Tehri dam, the "world's fifth and Asia's largest" dam, is an achievement, says the government. The closure of the tunnels, according to a THDC official, is a milestone in the history of the project.

On December 23, 2004 an advertisement appeared from the Information and Public Relations Department of the Uttaranchal government in several national dailies proclaiming the closure of the two tunnels of Tehri Dam (T3 and T4) as a "new chapter in development". It highlighted the salient features and benefits of the project, the status of the rehabilitation measures, and so on. It said the people of Tehri had made a great and unforgettable sacrifice that had enabled the construction of the dam. "The government was dedicated to the full and proper rehabilitation of the people of Tehri town and surrounding villages and also to the development of regions beyond the dam that are affected. In the first phase of rehabilitation, 3,251 urban families and 2,583 rural families have been rehabilitated. In the second phase, 2,604 families from the villages in full submergence and 3,663 families from the villages in partial submergence will be rehabilitated," it said.

According to government figures, of the total number of 2,409 affected rural families, 301 are yet to be rehabilitated in the first phase. In the second phase, of the 6,441 affected families, 2,303 are to be rehabilitated.

The people in Tehri town had been asked to vacate their homes by the end of March 2005,when the tunnels were to be closed. Central government set up a 10-member committee headed by Human Resource Development Minister to study the project and submit a report within three months. Its members included Director-General, Council for Scientific and Industrial Research; former

Secretary, Ministry of Water Resources; Director, Geological Survey of India; and Chairman, Central Pollution Control Board. The committee was asked to look into two aspects of the project: its safety (from the seismic point of view, in the context of the earthquake in Bhuj) and its impact on the purity of the Ganga. Its terms of reference did not include the issue of rehabilitation.

Sunderlal Bahuguna said that the agitators' main demand was just and fair rehabilitation. "Two acres of land is nothing without meeting the requirement of fodder, fuel and green vegetation," he said. Bahuguna also said that the THDC's claim that local manpower was being used in the project was false. According to him, mostly labourers from other parts of Country were being used and that many of them were bonded. A report brought out in November by the South Asia Network on Dams, River and People (SANDRP) gives a clear picture of the extent of rehabilitation. According to it, the government has not met even the basic requirements of a rehabilitation process. For instance, it does not have a database on the number of people affected and their incomes and livelihood patterns. Nor does it have a comprehensive policy that takes into account all aspects of displacement, the availability of suitable land and the need to settle village communities as social entities that are intact.

There is no master plan for resettlement and rehabilitation and the project-displaced people are not involved in the rehabilitation process, the report says and points out that the current package is inadequate for people to return to their original standards of living. While the Environmental Impact Assessment report of the project estimated the number of affected people at 97,000, the THDC puts the figure at 67,500. Only a limited extent of land is available for some project-affected persons and even that is of questionable quality, the report says. In some cases the land belonged to other communities.

The Silent valley project in late 1970's was another controversial project between Government and Environmentalists. The proposed project, now abandoned was to Dam the Kuntipuzha river in Kerala' s Palghat district. It flows through the valley, the Kuntipuzha drops 857m, making the valley an attractive site for generating electricity. The promoters of the project claimed that it would produce 240 MW of power, irrigate 10,000 hectares of land and provide 2000 jobs. On the other hand the environmentalists asserted that, Silent Valley, a home of few remaining rain forests in the Western Ghats, ought to remain prestine. They further conted that with over 900 species of flowering plants and ferns and several endangered species of animals and birds, Silent Valley was one of the world's richest biological and genetic heritage.

CHAPTER-V

Mineral Resources

Any naturally occurring concentration of a free element or compound of two or more elements in solid form is called a mineral deposit. Although a few minerals, such as gold and silver, occur as free elements, most are found as various compounds of only eight elements that make up 99.3% of earth's crust. A mineral deposit with a high enough concentration of at least one metallic element to permit it to be extracted and sold at a profit is called an ore or ore deposit. An ore that contains a relatively large concentration of a desired metallic element is called a high grade ore; one with a relatively low concentration is known as a low grade ore.

Table: showing the eight most dominant elements found on earth's crust

	Elements	Percentage by weight	Percentage by volume
1.	Oxygen	46.60	93.77
2.	Silicon	27.72	0.86
3.	Alluminium	8.13	0.47
4.	Iron	5.00	0.43
5.	Magnesium	2.09	0.29
6.	Calcium	3.63	1.03
7.	Sodium	2.83	1.32
8.	Potassium	2.59	1.83

USE AND EXPLOITATION OF MINERAL RESOURCES

The technologies to find and extract most of the minerals from the earth's crust, which are non-renewable is known to us.

These raw materials are converted into various items of everyday use, which are further reused, recycled or discarded. The demand for mineral commodities is increasing throughout the world because of the facts that the population as well as the per capita consumption are rising. Concentration of mineral resources formed millions of years ago are being exhausted in a very short span of time. Estimating how much of a particular non renewable mineral resource exist on earth and how much of it may be located and extracted at an affordable price is a complex and controversial process. The term total resources refers to the amount of a particular material that exists on earth. The future availability of a mineral resource depends not only on its actual or potential supply but also on how rapidly this supply is being depleted to the point where extracting and processing what remains is too costly. Five countries of the world- The USA, Canada, Australia, South Africa and Russia supply the world with most of 20 minerals that make up 98% of all non fuel minerals consumed in the world.

Depletion time is the length of time it takes to use up a certain proportion- usually 80% of the reserves of a mineral at a given rate of use. A traditional measure of the projected availability of nonrenewable resources is the reserve to production ratio- the number of years proven reserves of a particular nonrenewable mineral will last at current annual production rate. The shortest depletion time assumes no recycling or reuse and no increase in reserves. A longer depletion time assumes that recycling will stretch exiting reserves and the better mining technology, higher prices, and new discoveries will increase reserves. . While world population doubled between 1950 and 1993, global production of six key metals (aluminum, copper, lead, nickel, tin and zinc) increased more than 8 folds. During the same period, worlds reserves of copper increased 5-fold, lead almost 3-fold, zinc 4 fold and aluminum almost 9 fold.

Indian sub continent is rich in both, renewable and non renewable resources. As a matter of fact, the mineral exploitation in India has started late and there are encouraging indications of discovering many high cost, low volume and strategic minerals from this sub continent. The established mineral potential of India is so far been very limited, the reasons for this are not yet clearly understood, one of the reason may be the late start of exploration of resources in the Himalayan region. There is also the possibility that there may be a conceptual barrier in the search for mineral resources. The pre Cambrian of the Himalayan mobile belt show subdued base metal mineralization as in the case of crystalline of Kumaon, the salkhala rocks of Kashmir and Thimpu crystalline of Bhutan. Likewise with the characteristic volcano sedimentary formations, the phenerozoic of Himalayas and the lime stone deposits of Kumaon Himalayas. The dolomite and lime stone deposits of Bhutan and Arunachal Pradesh, the gypsum deposits of eastern Bhutan and the coal resources of Darjeeling, Sikkim, Bhutan and Arunachal Pradesh. The assumptions that these resources can not be saved are now proving wrong and it is seriously being realized that a severe shortage will develop if necessary attention is not paid for their conservation.

Table: showing the Important Mineral Reserves of India

S. No.	Mineral	Reserves (Tonnes)	States
1	Bauxite	252.53	Andhra Pradesh, Bihar, Gujrat, M.P., Karnataka, Goa, Jammu and Kashmir, Maharashtra, Orissa, Tamil Nadu, UP.
2	Gold	88	Karnataka, Andhra Pradesh
3	Iron Ore	1274	Andhra Pradesh, Kerala, Tamil Nadu, Karnataka, Goa, Maharashtra, MP, Orissa and Bihar
4	Lead	16.75	AP, Tamil Nadu, UP, Orissa, West Bengal, Sikkim, Gujrat, Rajasthan
5	Manganese	29.4	Maharashtra, Karnataka, Bihar, Orissa, MP Andhra Pradesh
6	Nickel	29.4	Maharashtra, Karnataka, Orissa, Bihar, Manipur, Nagaland, Orissa
7	Tungsten	2329	UP, West Bengal, Rajasthan, Maharashtra, Karnataka
8	Diamond	12.75	Andhra Pradesh, MP
9	Dolomite	496.7	Orissa, MP, Gujrat, Bihar, UP, West Bengal
10	Gypsum	23.90	Rajasthan, Tamil Nadu, Jammu and Kashmir, HP, Gujarat
11	Lime Stone	7645	MP, Tamil Nadu, AP, Bihar, Gujrat, Orissa, Rajasthan, Karnataka
12	Graphite	31.0	Kerala, Bihar, Orissa,, Rajasthan, Tamil Nadu, Andhra Pradesh

Source: Arora (1976)

ENVIRONMENTAL EFFECTS OF EXTRACTING AND USING MINERALS RESOURCES

According to an estimate made in 1960, a US citizen is supported by 9.4 tonnes of steel of which about 8% is used in cars, trucks and buses. Similarly they use 7.27kg of lead, 91 kg of clay, 227 kg of cement and 3.55 tonnes of stones, sand and gravel per capita per year. A relation between man and minerals has been given by Flawn in 1966. Man like an earthworm, burrows into the earth and runs over its surface, like a bird he brings materials from elsewhere to build his nest, and like a pack rat he accumulates quantities of trash. These activities have their ecological and environmental effects.

The highest danger of a high level of resource consumption may not be the exhaustion of resources but the damage that their extraction and processing do to the environment. Mining and processing minerals are among the most environmentally damaging of all human activities. After the mineral deposits are located in the earth crust, deep deposites are extracted by surface mining and shallow deposits by surface mining. The mining, processing and use of crusial resources requires a large amount of energy and often causes land disturbance, erosion, and air and water pollution.

Mining affect the environment in several ways and scarring and disruption of the land surface is the important one. The underground mines have a threat of collapse or subside, leading roads to buckle, house to tilt, rail road tracks to bend, sewer lines to crack, and disruption of ground water systems. The air is contaminated with dust and toxic substances, water pollution is a serious concern. Acid mine drainage occurs when aerobic bacteria produces sulfuric acid from iron sulfide minerals in spoil from coal mines and some ore mines. The rain water seeping through the mines carry the acid to nearby streams, destroying the aquatic life and contaminating surface water supplies.

The process of smelting the minerals in pure form from its ores or other elements causes further pollution in the environment. Smelters emit enormous quantities of air pollutants which include Sulpher di oxide, soot, tiny particles of arsenic, cadmium, lead and other toxic elements, which damage vegetation and soils in the surrounding areas. Smelting of copper and other nonferrous metals accounts for about 8% of human related emissions of sulpher dioxide each year. Similarly the land eroded, rivers dammed and land flopded to supply the electricity for metal processing, displacement of people and destruction of vegetation are some other important effects of over exploitation of minerals. Production of solid wates is another big environmental problem which is created due to extraction, use and through away of mineral resources.

CASE STUDY OF MINING AND QUERING IN RAJASTHAN

There are about 200 open cast mining and quarring centeres in a single district of Udaipur in Rajasthan and half of these are illegal and involved in stone mining including soapstone, building stone, rockphosphate and dolomite. The area has witnessed a large number of environmental problems due to these mines which are spread in an area of 15,000 hectares. Explosives used in blasting these mines not only causes noise pollution but also the air and water pollution in the area. The hills are devoid of any vegetation and suffer with large scale soil pollution. The Maton mines have polluted the Ahar river badly. The area already has very scarce water resources and people are compelled to use this polluted water for different purposes.

Similarly the mining activities in the Aravalli ranges have left many areas permanently infertile and barren. The precious wild life is under threat. The Sariska Tiger Reserve also lies in this area and has very rich wild life and also a large amount of mineral resources.

Mineral Resources Management

World's stock of minerals is considerably depleting because of their use in almost every walk of human activities. Time has come when appropriate attention is to be given and to understand the ill effects of reduced mineral supply for the well being of the human beings. There is an urgent need to call for the proper exploration and development of mineral resources in India. There are many organizations involved in the exploration, development and management of mineral resources in India. The important ones are listed as below:

(i) Geological Survey of India (GSI)

(ii) Indian Bureau of Mines

(iii) Public Sector Mining Undertakings

These are involved in the survey and search of minerals, conservation and development of mineral resources, providing technical assistance to mining industries, functioning as data bank of mines and minerals and advise Central and State Govts. on all aspects of mineral industry.

Mining the oceans

Ocean mineral resources are found in three areas: sea water, sediments and deposits on the shallow continental shelf and sediments and nodules on the deep ocean floor. Most of the chemicals found in sea water occur in such low concentrations that recovering them takes more energy and money, then they are worth. Only magnesium, bromine, and sodium chloride are abundant enough to be extracted profitably at current prices with existing technology.

Deposits of minerals along the continental shelf and near shore lines are already significant sources of sand, gravel, phosphates, sulfur, tin, copper, iron, tungsten, silver, titanium, platinum, and even diamond. Offshore wells also supply large amount of oil and natural gas. The deep ocean floor at various sites may be a future source of manganese and other metals. Manganese rich nodules are thought to cover 20% of the worlds ocean floors and have been found in large quantities at a few sites.

CHAPTER-VI

ENERGY AND FOOD RESOURCES

GROWING NEED FOR ENERGY

Energy is required in every part of our life, to keep our hearts beating, to think, to breath, to cook food, to travel by any method, and to warm or cool to our buildings in which we live. Energy powers the factories, cars, bulldozers, airplanes and electric motors Large amount of energy is also used to extract metals from ores and to manufacture the fertilizers, pesticides, plastics and many other products of every day use. We use energy to perform certain tasks where a certain amount of energy is essential. With the demand of growing population, the world is already facing a further deficit in energy resources.Our life style is changing very fast and from a simple way of life we are shifting to a luxurious life style.The countries like America, with only 4.7% of world's population is using 25% of the commercial energy. It is also the largest waster of energy. Average per capita energy use in the United States is almost twice that of Japan and many other developing countries.

The earlier agricultural society used the energy of the solar energy which was captured by the green plants for production of food and through the labour of the domesticated animals. Advanced agricultural man produces surplus food through energy-intensive crops, which enables the society to support the people who are not farmers. The energy use has further increased in the advanced societies which are basically dependent on large industries. Here the solar energy is being supplemented with the use of fossil fuels and the level of energy consumption has tremendously increased all through the industrial revolution. The energy used by the modern man is more than 200 times the basic need of hunter and food gatherer societies. Future projections made for energy requirements have always fallen short. However, it is certain that the energy requirements throughout the world shall increase in times to come and an urgent need to look for the alternative sources of energy will be inevitable.

Table; Per capita energy use and GNP of some countries (World Resource Institute)

Country	Gross National Product $ per capita	Per capita energy consumptionin GJ
Switzerland	35,000	100
Japan	32,000	100
Denmark	25,000	100
Germany	20,000	150
Sweden	25,000	250
Norway	26,000	310
USA	22,000	350
UK	18,000	150
Argentina	8,000	50
Egypt	3000	20
India	2000	10
China	3000	15
Greece	6000	100

RENEWABLE AND NON-RENEWABLE SOURCES OF ENERGY

Energy is the capacity to do work. It comes in many forms like light, heat, electicity, chemical energy, from coal, moving matter such as water, wind and also nuclear energy emitted from the nuclei of certain isotopes. An energy source can be one which has the capacity to provide adequate amount of energy that can be used for a long time. The different sources of energy can be grouped in to two categories:

1. Renewable Sources of Energy

These include solar energy, wind energy, tidal energy, hydropower, biomass energy, geo-thermal energy, which are also known as non-conventional sources of energy. The other renewable sources of energy include food, fodder, fibre which provide energy to human beings for different activities.

2. Non-Renewable Sources of Energy

These include the sources like coal, petroleum, natural gas and nuclear fuels which are present in the nature in limited quantities and can not be replenished when exhausted.

ALTERNATE SOURCES OF ENERGY

The sources of energy which are not in regular use, however, which can be potential sources of energy in future can be said as the alternative source of energy like biomass energy, alcoholic energy, geothermal energy and energy from petro-crops like Jatropa.

SOLAR ENERGY

This is the energy derived from the sun. The sun is essentially a thermo nuclear reactor and is almost inexhaustible source of energy. Sun's heat energy is used directly by plants, which are the primary producers and this energy is then converted into chemical energy through the process of photosynthesis, which is utilized by secondary and tertiary producers in the food chain. Sun's heat energy is being used directly in solar cookers to cook food or in solar heater to heat water. The sun's energy can be concentrated by big concave reflectors and used to boil water to obtain steams and produce electricity.

Solar energy is kinetic energy radiation. Since using solar energy will not deplete the sun in any way, astronomers estimate that the sun will burn for another several billion years-it is, therefore, commonly referred as renewable energy.

It is calculated that a very small percentage of solar energy would be more than enough to meet all our energy demands for transportation, industry and residential purposes. Not only now but for the foreseeable future.

The abundance of solar energy, however, stems from the fact that it falls over the entire earth; so where it is particularly intense. Thus the hurdles to be overcome in using solar energy involve, absorbing it in a broad area, and concentrating and converting it to forms that can power our transportation, residential and industrial needs. The difficulties or cost of overcoming these hurdles have given solar energy the representation of being impractical, at least for the present.

The key to using solar energy is in applying it to situations where the costs are minimal. As technological developments lower the costs, and traditional energy sources become more expensive, solar energy may be extended to additional applications. High energy may be captured and used directly as it comes from the sun; this is referred to as direct solar energy. Additionally, in the biosphere, winds, the water cycles, and the production of biomass are driven by solar energy. Thus taking energy from these sources is, in fact, indirect solar energy.

A passive solar heating system captures sun light directly within a structure and converts it into low temperature heat for space heating. Energy efficient windows, green houses and suspense face the sun to collect solar energy by direct gain. Thermal mass (heating storing capacity) in the form of walls and floors of concrete, brick, stone, salt treated timber or tiles store much of the collected solar energy as heat and release it slowly throughout the day and night.

In an active solar heating system, specially designed collectors absorb solar energy, and a fan or a pump is used to supply part of a building's space heating needs. Several connected collectors are usually mounted on a portion of the roof with an unobstructed exposure to the sun. Some of the heat is used directly, and the rest can be stored in insulated tanks containing rocks, water or heat absorbing chemicals for release latter as needed. Active solar collectors can also supply hot water. About 12% houses in Japan, 37% in Australia and 83% in Israel also use such systems. Currently about 1.3 million active solar hot water systems and 200,000 active solar space heating systems are in US homes.

Solar Radiations and its spectral characteristics

The wave concept of solar radiations explains how the energy propagates, but this can only be detected when it interacts with matter. In the interaction these radiations behaves as though it consists of many individual bodies called photons.

Frequency (ν): It is the number of wave crests passing a given point in a specified point of time. Frequency was formerly expressed as cycles per second, today it is expressed in terms of hertz, the unit for a frequency of one cycle per second.

Wavelength (λ): Wavelength is the distance from any point on one cycle or wave to the same position on the next cycle or wave. The electromagnetic spectrum ranges from the very short wavelengths of the gama rays to the long wave lengths of the radio region.

The following table shows spectral regions of the solar radiations:

Gama rays	less than 0.03 nm
X-ray	0.03 to 3.0 nm
UV rays	0.03 to 0.4 mm
Photographic UV band	0.3 to 0.4 mm
Visible	0.4 to o.7 mm
Infra red	0.7 to 100 mm
Reflected IR band	0.7 to 3.0 mm
Thermal IR band	3 to 5 mm and 8 to 14 mm
Microwave	0.1 to 30 cm
Radar	0.1 to 30 cm
Radio	> 30 cm.

The greatest amount of output of energy is in the visible light part of the spectrum.

The photovoltaic cell or solar cell

Solar cell, technically known as photovoltaic cell, is the device which enables a direct conversion of light energy to electrical energy. Generally constructed from a sandwich of two thin layers of silica which are treated such that light striking one causes electrons to fall to other. Solar energy can be converted directly into electrical energy by photovoltaic (PV) cell, commonly called solar cells. Sun light falling on a solar cell- a transparent water, thinner than a sheet of paper- releases a flow of electrons, creating an electrical current. Because a single solar cell produces only a very small amount of electricity, many cells are wired together in a panel providing 30-100 watts. Several panels are in turn wired together and mounted on a roof and produce electricity for a home or building. The DC electricity produced can be stored in batteries and used directly and converted to conventional AC electricity.

Solar cells are reliable and quite, have no moving parts and should last for 30 years or more if incased in glass or plastic. The cells can be installed quickly and easily and expended or moved as needed, maintenance consists of occasional washing to keep dirt from blocking the sun rays. Arrays of cells can be located in deserts and marginal lands, alongside interstate highways, in yards and on roof tops. In 1995, Enron, a major natural gas company, announced plans for a solar photovoltaic power plant backed up by energy efficient turbines burning natural gas that was expected to generate 100 megawatts of electricity.

WIND ENERGY

For thousands of years wind energy has been exploited but its re-emergence as one of the most cost effective renewable source for generation of grid quality electricity is of relatively recent origin. India has not only been quick to make a foray into this area but has also made a mark as one of the top ranking countries in the world in wind power generation. With an installed capacity of 1080 MW of wind power, India now ranks 5th in the world after Germany, US, Denmark and Spain. The most prominent feature of wind climatology in India is the monsoon circulation. The wind speed and power density are influenced by the strong monsoon, starting in May-Jun, when cool, humid air moves towards the land and by the weaker north-east winter monsoon, starting in October, when cool dry air moves towards the ocean. From March to August, the winds are uniformly strong over the entire Indian Peninsula.

GEOTHERMAL ENERGY

This is one of the most important natural resources, the real utilization of which is yet to be made. Because of the energy released in the decay of naturally occurring radioactive materials, the earth's interior is a huge ball of molten rocks, which occasionally erupt to the surface in volcanoes. This heat of the earth's interior is referred to as geothermal energy. It can be obtained at a practical cost and is virtually infinite and ever lasting. Some geothermal energy may be obtained where natural water comes in contact with hot rocks. Such heated water may come to the surface in natural steam vents. The steam is used to drive turbo generators. As of 1988, there were 5000 megawatts of such geothermal facilities installed, about half of them in California and others in Philippines, Itly, Mexico and Japan. This amount could be increased by several folds in the coming years. Large scale development of geothermal power presents many problems. Hot streams and water brought to the surface are frequently heavily laden with salts and other contaminants, particularly sulphar compounds. These contaminating compounds are highly corrosive to turbines and other equipment, and they result in both air and water pollution as they are finally released in to the environment. Sulfur pollution from a geothermal plant may be equivalent to that from a plant burning high sulfur coal, and the hot brines released into streams and rivers may be ecologically disastrous.

The real potential for geothermal energy lies in development of hot dry rock. Hot dry rock can be found any where by drilling deep down. In theory, two parallel holes can be drilled in to hot rocks and then fractures created between the two holes. Water forced down in one hole will be heated as it seeps through the fractures and will come up the other hole as steam to drive turbine. Recycling the water would avoid the problem of pollution.

TIDAL POWER

A great deal of energy is inherent in the twice daily rise and fall of the tides, and many imaginative schemes have been proposed for capturing this eternal, pollution free source of energy. The most straightforward is to build a dam across the mouth of a bay and mount turbine in the structure. The incoming tides flowing through the turbines generate power. As the tide shifts, the blades are reversed so the out flowing water continues to generate power. Russia and France are the two countries who have established two such plants. But tidal power is not without adverse environmental impacts. In addition to loss of unique aesthetic and recreational pleasure in these areas, there would be far reaching environmental impacts due to the dam's trapping the sediments, impending of the migration of the marine organisms and from changing circulation and mixing of fresh water.

NON- RENEWABLE ENERGY RESOURCES

FOSSIL FUELS

Since the origin of coal, crude oil and natural gas is a biological production, they are frequently called fossil fuels. While fossil fuels are formed by ongoing, natural process, two factors preclude any notion of their being replenished as we use them. First, biological conditions on the earth have changed so that significant accumulation of organic matter no longer occurs and second. We are using them far faster than they ever formed. It is estimated that about 1000 years of natural production was required to produce the crude oil that we now use each day. Because supplies are finite, therefore, sooner or later we will run out of these fuels. Fossil fuels are defined as any recognizable organic structure or impression of such structure preserved from prehistoric times. It is one of the most important source of energy in the form of coal, oil and natural gas.

(a) Petroleum

Deposits of petroleum or the crude oil and natural gas are trapped together deep underground, beneath a dome of impermeable cap rock and above a lower dome of sedimentry rock such as shale, with the natural gas lying above the crude oil. Crude oil is also found beneath the sea floor and, once it is removed from its reservoirs or deposits, most crude oil is sent by pipeline to a refinery. There it is distilled to separate it into component chemicals, which boil at different temperatures and are removed from various levels of giant distillation columns. Some of these chemicals are known as petrochemicals and are sent to petrochemical plants for use as raw materials in the manufacture of most industrial chemicals, fertilizers, pesticides, plastics, synthetic fibres, paints, medicines and numerous other products.

Petroleum is an oily, inflammable liquid made up mostly of hydro-carbons containing only hydrogen and carbon. The hydro-carbon content of petroleum ranges from 50-98%. The rest is chiefly made up of Oxygen, Nitrogen and / or sulpher. The greatest amount of petroleum is consumed as gasoline or diesel for driving of cars, buses and trains of the world. The rate of petroleum consumption is increasing annually at the rate of 6%. Petroleum, one of the most potent factor in the establishment of huge power stations, mammoth factories and various fast moving means of transportation, is soon going to exhaust. The remaining supply of oil should be reserved for lubrication, aircraft and petrochemicals. Its wasteful use in space heating or cooling should be banned. This will save 12.5 million barrels of petroleum in USA alone. Industrialized agriculture requires more petroleum than any other industry.

Originally it was estimated that close to 1250 million barrels of oil were deposited in the earth and out of this, approximately 100 million barrels were extracted by the end of 1960. South west Asia is the biggest oil producing region in the world. India is exploiting oil and natural gas from Gujarat, Arabian Sea, Assam and Godawari basin.

In addition to the above fossil fuels, oil shale and tar sands are believed to contain more than 3000 billion barrels and 7000 billion barrels of oil respectively. These can be recovered from the ground and can be potential sources of energy.

Oil Shale

These are compact sedimentary rocks from which petroleum is obtained by the process of destructive distillation. It is crushed, heated in a furnace where air is not allowed to admit. A very high temperature is kept so that chemical decomposition may take place. Thus oil, gases, water and solutions of other acids and substances are produced. It takes the energy equivalent of about one-third of a barrel of conventional crude oil to mine, retort and purify one barrel of shale oil, thus sharply reducing its net useful energy yield compared to conventional oil. As a result the price of extracting and processing of this oil is high. Environmental problems may also limit shale oil production. Shale oil production requires large amounts of water, which is scarce in the semi arid areas where the richest deposits are found. During the processing of this oil a large amount of carbon-dioxide and other possibly cancer causing substances are also released into the atmosphere.

Tar Sand

These are deposits of a mixture of fine clay, sand, water, and variable amounts of bitumen, a black, high sulfer, tar like heavy oil.Tar sands of Northern parts of Canada's Alberta Province, contain one of the largest deposits of petroleum in the world. Tar sand is mined and then washed to remove the tar from the sand. Tar thus obtained is then heated and cracked into simpler molecules which are upgraded and landed to produce synthetic oils.

Conservation is often promoted in terms of turning off lights, turning down thermostats and, car pooling. These activities can, and do produce immediate fuel saving and they have helped get us through periods of limited supplies. However, such activities have distinct limitations. The actual savings are small , only up to 2 to 4% of overall energy use, and they do entail some inconvenience. The real focus and potential of conservation is in the development of systems that use energy more efficiently so that the same or even improved transportation, lighting, heating, work, comfort etc. are achieved but less energy is consumed in the process. Petroleum is a non-renewable energy source, and if all the estimated supply of crude oil are discovered and developed and sold at a price of USD 95 a barrel, 80% of these reserves would be depleted by 2076 at 1984 usage rates and by 2037 if the usage increase by 2% a year. Thus relatively little of the world's crude oil is likely to remain by 2059, bicentennial of the world's first oil well.

Because of its several important advantages petroleum has a wide spread use. It is still a cheaper source of energy, can easily be transported and put into several usages such as –it can be burned to propel vehicles, provide low temperature heating of water and buildings, provide high temperature heat for industrial processes and production of electricity. Its disadvantages include release of carbon di-oxide on its burning, oil spills cause water pollution and also contaminate ground water.

(b) Natural gas

Natural gas consists of 50 to 90% methane and smaller amounts of heavier gaseous hydrocarbon compounds such as propane and butane. Although most natural gas used so far lies above deposits of crude oil, it is also found by itself in other underground deposits. Conventional supplies of natural gas are projected to last somewhat longer than those of crude oil. Between 1973 and 1984, proven reserves of natural gas doubled, whereas those of petroleum and coal have not risen. Much of this increase has come from large discoveries in the Russia which now has 40% of world's proven reserves and is the world's largest natural gas producer. Other countries with large natural gas reserves include Iran(14%), USA (6%), Qatar(4%), Saudi Arabia(3%) and Nigeria(3%). The world's identified reserves of natural gas are projected to last until 2033 at 1984 usage rates and to 2018 if usage increases by 2% a year.

(c) Coal

Natural coal is a solid, formed in several stages as the remains of plants are subjected to intense heat and pressure over millions of years. It is mostly carbon with varying amounts of water and small amounts of nitrogen and sulfur. Coal with high sulfur and nitrogen contents produces high emissions of sulfur dioxide and nitric oxides into the atmosphere when it is burned.

There are several major kinds of coal, each with a different carbon content, moisture content, sulfur content and fuel value.

(i) Anthracite, also known as hard coal, burns cleaner with less smoke than other types, but it is not common and is usually more expensive. It has a shiny black color and is some time made into costume jewelry. It is highly valued as domestic heating fuel.

(ii) Bituminous coal, or soft coal, is much more abundant but usually has a higher sulfur content. It has a good storage quality.

(iii) Sub-bituminous coal and ignite are generally low in sulfur but also have a low heat content and a high moisture content and is burned for fuel in areas such as Ireland where supplies are plentiful. It is good source of industrial gases and carbon. It has lowest percentage of fixed carbon and highest content of volatile matter.

Coal is the world's most abundant fossil fuel. About 68% of the world's proven coal reserves and 85% of estimated undiscovered coal deposits are located in three countries: USA(29%), USSR(28%) and China(11%). These countries also account for about 60% of present total world coal production. India accounts for about 0.8% of the total geological reserves of coal and 5.7% of the proved reserves of coal in the world. There are 201 tones of coal reserves per person in India as compared to 13,742 tones in USA, 23,112 tonnes in former USSR and 142.35 tones in China.

Coal Reserves In India

As a result of explorations carried out down to a depth of 1200m by GSI and other agencies, a cumulative total of 220.98 billion tones of coal reserves have been established in the country. The state wise distribution of these reserves are as follows:

	State	Coal Reserves in Million Tones
1.	Andhra Pradesh	15462
2.	Arunachal Pradesh	90
3.	Assam	320
4.	Jharkhand	69175
5.	M.P/ Chattishgarh	49543
6.	Maharashtra	7362
7.	Meghalaya	459
8.	Nagaland	20
9.	Orissa	51571
10.	U. P.	1062
11.	W. B.	25919
	Total	220983

The type wise and category wise reserves are as follows

Type of Coal		Reserves in Million Tones
(A)	Coking	
	Prime Coking	5313
	Medium Coking	23626
	Semi Coking	1610
(B)	Non Coking	190434
	Total	220983

Coal Conservation In India

Conservation of coal enjoins maximum recovery of in-situ reserves of coal. Coal deposits in India occur mostly in thick seams and at shallow depths. These aspects are taken in to account during mine planning and operation for ensuring maximum recovery.

Mechanized opencast mining in India is one of the very important technologies of coal production of thick seam from shallow depths. The percentage recovery by this method is upto 80% to 90% of the in-situ coal reserves. The coal production from open cast method in Indian mining is more than 80% of total production. This trend is likely to continue in the near future. This thick seam deposit earlier developed in board and pillar method or other methods of underground mining which has been standing on pillars for long in absence of the suitable technology of extraction have now in many cases become extractable by open cast mining equipment of suitable type. Mines having difficult geo-mining conditions like steep and irrigular coal seam deposits, special underground mining methods such as, shield method, flexible roofing method, bhaska method, chabmer method, Tipong method, scraper-assisted chamber method etc. have been tried out of which some methods have been successfully adopted.

Sand Stowing

Sand stowing in underground mines is yet another effective means of coal conservation, which is widely used for extraction of coal pillars from underground coal seams and coal seams lying underneath built up areas, such as important surface structures, railway lines, roads, rivers, nallahs, Jores etc., which otherwise would have resulted in locking of coal pillars. However, the economic viability of underground mining with sand stowing is getting bleak with increasing cost of stowing.

Fossil fuel management

Fossil fuels are non-renewable sources of energy and get destroyed during the process of combustion. Improvement in their efficiency to obtain commercial energy is very important. Their proper management can be achieved by practicing the following:

(i) Reducing the demand of non-renewable resources of energfy.

(ii) Increasing the use of non-conventional sources of energy for all purposes.

(iii) Intensifying surveys and search work to find out the new reserves.

(iv) Carrying out more research aiming to improve energy efficiencies in different sectors.

(v) Creating awareness among public at large regarding efficient use and conservation measures.

(vi) Reducing loss of fuels during production and transportation of such fuels.

NUCLEAR ENERGY

From the time fossil fuels were first used, it was being realized that they would not last forever. Sooner or later, other energy sources would be needed. It has been anticipated that nuclear power could produce electricity in such large amounts and so cheaply that we would phase in to an economy in which electricity would take over virtually all functions, including the generation of other fuels, at nominal cost. The release of nuclear energy is a completely different phenomenon from the burning of fuels or other chemical reactions. Nuclear energy, however, does involve changing atoms through one of the two basic processes, fission or fusion. **In the process of fission, a large atom of one element is split, resulting in**

two smaller atoms of different elements. While in the process of fusion, two small atoms combine to form a large atom of different element. In both fission and fusion, the mass of the products is less than the mass of the starting material, and the lost mass is converted into energy. The amount of energy released by this conversion is tremendous, however, the environmental pollution risk is very high.

All current nuclear power plants utilize the fissioning of Uranium atoms, specifically U235. Uranium is an element that occurs naturally in various minerals in the earth's crust. It exists in two primary forms or isotopes, Uranium-238 and Uranium-235. Occasionally, a uranium-235 atom fissions spontaneously, but it can be triggered to fission if it is struck by neutron. This is the key to promoting nuclear reaction. When one U-235 atom splits , it ejects two or three neutrons in addition to releasing energy. If one of these neutrons strikes another U-235 atom, fission occurs again thus releasing more energy and more neutrons, which repeat the process. France , Germany and UK are the highest producer of atomic energy. A tone of U235 would provide as much energy as 3 million tones of coal or 12 million barrels of oil. The nuclear power plant in the world provide only 3% of the worlds electricity energy. India has established five power plants, 3 of which are located in Tarapur, Kota and in Tamil Nadu. And contributing 2.3% of the total electricity production in India.

FOOD RESOURCES

The term food has been defined as ***anything that is able to satisfy the apetite and to meet the physiological needs for growth, to maintain all body processes and to supply essential energy required for maintaining body temperature and activity.*** The most important components of food include carbohydrate, proteins, fats, minerals, vitamins and water. Of earth's more than 32000 plants with edible parts, we eat only about 30. 90% of our food is supplied by just 15 plants and 8 animal species. Crops like wheat, rice, corn and potato make up more of the world's total food production than all other combined. Grains provide us about half the world's calories, with two out of three people eating mainly a vegetarian diet. With the rise in income , people consume more grain indirectly in the form of meat, eggs, milk, cheese and other products of domesticated livestock. Two types of agricultural systems have been identified which are industrialized and traditional. The industrialized agriculture uses large amounts of fossil fuel energy, water, commercial fertilizers and pesticides to produce large quantities of one crop or animal for sale. The industrialized farmers use following in order to grow their food resources:

(i) Grow surplus food for sale by investing a large amount of money.

(ii) Use hybrid seeds of a single crop variety on a large field.

(iii) Good and high-tech equipments are used for agriculture.

(iv) Yields are increased with commercial fertilizers and irrigation.

(v) All shorts of chemicals are used as pesticides.

(vi) Fossil fuels powered machinery is used to cultivate the crops.

(vii) Produce fatty meat that most consumers like.

(viii) Use feedlots to fatten hundreds of live stocks in a small space.

The traditional farmers in turn practice the following:

(i) Grow enough food to feed their families and invest little money

(ii) Plant a mix of naturally available crop seeds on a small plot

(iii) Use simple equipments costing relatively little

(iv) Increase crop yields by using naturally available water and organic fertilizers.

(v) Cultivate by hand or with the help from draft animals.

(vi) Use natural grasslands and forests as source of food and water for small groups of live stock.

(vii) Produce lean meat which is more healthful than fatty meat

WORLD FOOD PROBLEMS

During a span of 40 years between 1950 and 1990 the world grain production has increased more than three times while the per capita production has increased by 50%. During the same period average food prices adjusted for inflation dropped by 25%, and the amount of food traded in the world market increased by 4 times. However, population growth has outstripped food production and according to a report food production has lagged behind population growth in 69 low developed countries. The annual rate of increase in global food production dropped from a high of 3.5% between 1996 and 1976 to about 2.2% between 1980 and 1990. Similarly in 22 African countries per capita food production has dropped 28% since 1960 and may drop another 30% during the next 25 years. A large number of countries are regularly importing the food from United States, Canada, Australia, Argentina and western Europe.

India, known as an agricultural nation, is the third largest producer of staple crops, but an estimated 300million Indian are still under nourished. With its huge population, the food problems are keep on increasing, though we have only half as much land as USA.

Table : Showing food requirements for average human beings

Average adult	1700 Cal/ day at rest	Women 2200 Cal/day at work	Men 2800 Cal/day at work
Protein	70 gm	Vitamin A- 4700 IU	
Calcium	1 gm	Vitamin B1 -1.6 mg	
Iron	12 gm	Vitamin B2 – 2.3 mg	
		Niacin – 16.0 mg Vitamin C – 70.0 mg	

To maintain good health and diseses resistance, people need not only a certain number of calories but also food with the proper amount of protein, carbohydrates, fats, vitamins and minerals. Population, which is forced to live on a low protein, high starch diet of grains such as wheat, riceand corn suffer with malnutrition Or difficiencies of protein and other important nutrients. According to a report of WHO about 1.3 billion people are underfed. Each year 40 million people and half of them are children under the age of 5 years die prematurely from malnutrition. The people who remain chronically undernourished and malnourished are prone to diseases and too week to work productively.

There is also concern over possible harmful effects of chemicals which are known as food additives, which are added to processed foods for sale in grocery stores and restaurants. The addatives retard spoilage, enhance flavour, color and texture. The presence of synthetic chemical additives does not necessarily mean that a food is harmful.

CHANGES DUE TO OVER GARZING AND AGRICULTURE

Overgrazing

In every country the population of cattle is maintained for the milk and Agricultural Uses: Cows, goats and buffaloes are reared for their milk in different villages. Some villages also depend upon Gujars or nomadic/semi nomadic tribes for their milk requirement. Bullocks are used for ploughing the field; for obtaining manure: Animal dung has a prominent place in hill agriculture where chemical manures are not in much use. Each farmer likes to keep about 10 animals per hectare of cultivation in the hilly areas. Many farmers also keep one or two stall-fed buffaloes, but stall-feeding of cows is rare. Usually, animals kept exclusively for this reason are allowed to graze on the pastures during the day and they are kept in enclosures during the night; and for wool and meat: Some nomadic and semi-nomadic tribes in the hills and foot-hill zones rear goats, sheep and yaks for their hair, which is used to make woollen garments, blankets, carpets etc. Overgrazing by cattle is one of the greatest factors in the degradation of pasture lands particularly in the Himalayan areas. Over the years, with the increase in cattle population, there has been a rapid degradation of the forest environment. In J & K, after 1970 the cattle population nearly doubled. 20% of the total cattle population of the country is in the Himalayan region, which has about 18 percent of the forest area of the country.

The over garzing leads to land degradation as it removes the vegetal cover from over the soil, the roots cannot hold the soil properly, loss of adequate soil moisture and infiltration capacity of soil also takes place. Over grazing also inhances the soil erosion by wind and water as the soil becomes exposed to these factors and gets eroded. Due to over grazing the composition of plant population and their regeneration capacity also gets affected. When the heavy grazing on original grasslands takes place , the hoofs of the cattle destroy the roots of the plants and chances of their regeneration become almost nil. As a result the vast areas are invaded by easily growing weeds.

Agriculture

Primitive societies used to obtain their food from hunting anf food gathering in the forests. Even today some tribal communities in MP and other states in India depend on hunting and food gathering. With the rapid development in the various equipments, synthetic fertilizers and pesticides modern civilization obtains its food from the cultivated plants and domesticated animals. The type of agriculture practiced now a days is very different from the traditional practices and outputs in terms of yield have increased many folds and so have increased the environmental impacts. The important impacts of modern agriculture are as follows:

(I) Converting forest land into short lived pastures for cattle to raise cattle products like milk and meat has destroyed large areas thus reduced the natural repositories of our biodiversity.

(II) About 15% methane released in to the atmosphere comes from the belching by cattle.

(III) Fertilizer problem: Addition of too many and too much of fertilizers has changed the chemistry of land. It leads to soil and water pollution and according to a report of US Public Health Service, water containing more than 10 mg of nitrate per liter is unsafe and this nitrate come in water from the agriculture fields.

(IV) Addition of fertilizers changes the pH of soil which changes the solubility of different minerals in the soil water, thus making them more or less available to plants.

(V) Pesticide Problems: Indiscriminate Use of pesticides is causing a threat to human health apart from destroying the non-target pests and vegetation. Pesticides pass in to food chain directly or indirectly thus cause a health hazard.

(VI) Pesticides are also converted into chemicals which are potentially more harmful than the original form e.g., heptachlor, a pesticide, is converted into a stable substance which is 22 times more toxic than the original pesticide, when exposed to sun light.

(VII) Development of resistant variety of pests and weeds is another problem associated with the large scale use of pesticides. Following the bulk application of chemical pesticides after the second world war, the number of pest species with resistance power against the original pesticides, increased dramatically.

(VIII) Salinity Problem: According to an estimate about one third of the cultivable land of the world is affected by salts. The scene is no different for India also where about seven million hectares of land is estimated to be salt affected and is either saline or sodic. The soil which are saline have a large accumulation of soluble salts like sodium chloride,sodium sulphate, calcium chloride, magnecium chloride etc.in the soil Whereas the sodic soils have carbonates and bi-carbonates of sodium. Excessive irrigation is the major cause of salinization of soil Ground water unlike rain water contains dissolved salts and under dry conditions, the water evaporates leaving behind salts in the soil. The salinity is responsible for the stunted plant growth and low crop yield.

(IX) Water logging: Over irrigatiobn of agriculture fields causes the problem of water logging. In the states of Punjab and Haryana large areas have become water logged where adequate water supply from canals or tubewells have encouraged the farmers to go for excessive irrigation of crops. In water logged areas the mechanical strength of soil declines, the crop plants get logged and ultimately the crop yield falls.

CASE STUDIES

It has been reported that several villages in the State of Haryana which are lying in the command area of western Yamuna Canal are suffering from destructive saline efflorescence, Similarly the floods in Punjab during 1950, 52, and 1954 -56 resulted in aggravated water logging with serious drainage problems. The Indira Gandhi Canal Project has also caused a water logging problem in the state of Rajasthan changing it from water starved land to water soaked wasteland.

CHAPTER-VII

LAND RESOURCES

LAND AS A RESOURCE

The term land refers to all features of the natural environment of a part of the earth's surface, to the extent that they exert a significant influence on its potential for use by man. Thus land covers geology, land forms and soils but also the complete, vegetation and fauna, including pests and diseases. Where any forest stands are present, these form part of the land. Land is the most important natural resource and is to be utilized in accordance with its suitability and ecological capabilities, which depends on geological and biological factors. However, if unbearable pressures such as of an explosively growing population are brought to bear on the land and to over use it and over exploit its limited resources, the ecological balance is upset. Soil can be described as a mixture of rock and mineral material with organic matter. It can be better described as a three phase system composed of solid , liquid and gas which interact with each other. Soil is the net result of the interaction of climate and organisms, especially plants on the earth's crust.

Soil provides nutrients, support and water for the growth of the plants and trees which provide food, fodder, fuel and other products to the man and his subjects. Distribution of flora and fauna is largely dependent on types of soil. Soil plays an essential role in the cycling of various types of nutrients and gases in the atmosphere with the help of decomposer organisms which are present in the soil.

Due to increase in population, the land resources are under great pressure. In the year 1901 the population of India was about 238 million which has become one billion and is being supported by the same land area of 32,88,000 km2 . The per capita land resource in India is less than 0.4 ha while it is 0.9 ha in China and 8.4ha in the erstwhile USSR. In India about 44% of land is put under the agriculture , 22% is covered with forests, 5% is used for pasture and grazing fields, 17% for housing, industries, roads and other various developmental activities and about 14 % is lying as the barren land of no use. The rational use of land resources is possible only by adopting an integrated land use policy, which involves prevention of land misuse, and reclamation of degraded and underutilized land.

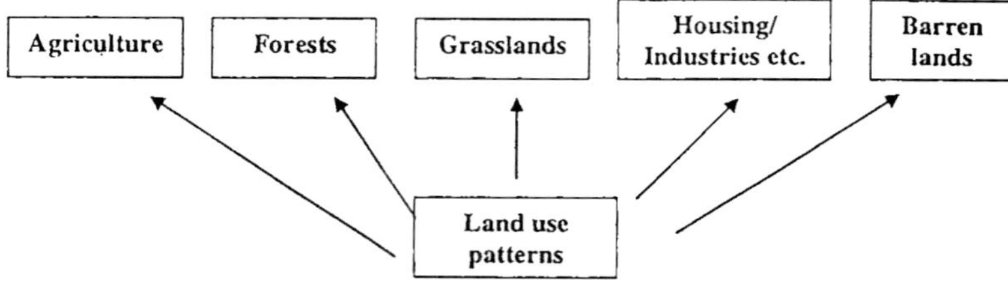

Land use patterns

The term land use and kind of land use are employed in a general sense to refer to any form of use of land by man. Each kind of land use has a set of environmental conditions which are favorable or adverse to its practices. The land use requirements are the conditions of the land necessary for the successful and sustained practice of a given land utilization type. Examples are suitable temperatures and soil rooting conditions to permit tree growth, or terrain conditions favorable for mechanized forestry operation, such as gentle slopes and absence of rocky substratum.

Land units are delineated areas of land with specified environmental conditions, employed as the basis for land evaluation. A land unit does not refer to any particular kind or scale of mapped areas, but is a term of convenience to cover any such area that is selected for evaluation.

LAND DEGRADATION

Land degradation refers to the process where loss of soil productivity takes place and the capacity of soil to produce goods or services are lost with the passage of time. The area of land surface which is potentially available to grow crops is declining every year. Large areas which were once used for cultivation purposes have already become practically unusable by erosion. Soil crosion is the movement of soil components, especially surface-litter and top soil, from one place to another. The important causes for this movement are water and wind. Some soil erosion is natural, but the roots of plants generally hold the soil. The other main deteriorater of land is due to fire. The burning of agricultural land and forests removes practically the entire supply of nitrogen. Apart from the loss of different nutrients, there will be a drop in the water supply and a change in the microclimate. In the deforested areas, there will be flooding during the rainy period while such areas will experience draughts during dry season.

According a report, the top soil is eroding faster than it forms on about one third of the world's cropland. In some countries more than half land is affected, which include, Nepal (95%), Peru (95%), Turkey (95%), Lesotho (88%), Medagaskar (979%) and Ethopia (5c%). A study carried out by World Resource Institute in 1992 says that soil on more than 12 million square kilometers of land that is an area of the size of India and China combined- had been seriously eroded since 1945. It also found that 89,000 square kilometers of land scattered across the globe was too eroded to grow crops any more. Annual erosion rates of soil throughout the world are 20-100 times the natural renewal rate.

In mountainous areas such as in Himalayan region, farmers traditionally built elaborate system of terraces, Terracing allows them to cultivate steeply sloping land that otherwise quickly loose its top soil. However, with the passages of time, these slopes are being farmed without terraces, leaving the soil too poor after 10-40 years to grow crops or generate new forestsThe resultant loss of protective vegetation and top soil also greatly intensifies flooding below these watersheds.

DESERTIFICATION

It is the process where the productivity potential of arid or semi-arid land falls by 10% or more and is caused mostly by human activities. It is a serious and growing problem in many parts of the world. The regions most affected by desertification are all cattle producing areas. Practices that leave topsoil vulnerable to erosion and drying include overgrazing, deforestation without reforestation, surface mining without land reclamation, irrigation techniques that lead to increased soil erosion, salt build up and water logged soil. These destructive practices are linked to rapid population growth, high human and livestock densities, poverty and poor land management.

It is estimated that 810 million hectare have been desertified during last 50 years. The UN Environment Program estimates that world wide 63% of rangeland, 60% of rain fed crop land and 30 % of irrigated croplands are threatened by desertification. Every year an estimated 60,000 square kilometers of new desert are formed, and another 210,000 square kilometers loose so much of top soil and fertility that they are no longer able to be used for farming and grazing.

LAND SLIDES

Land slides and related phenomenas cause substantial damage and loss of land . Land slides refer to a rapid down slope movement of rocks or soil as a more or less coherent mass. It is also used as a comprehensive term for any type of down slope movement of earth materials. Climate and vegetation can influence the type of landslides pr other downslope movement that occurs on a particular slope.Climate controls the nature and extent of precipitation and thus the moisture content of slope materials. The role of vegetation in an area is a function of several factors, including climate, soil type, topography and fire history each of which influences what happens on the slope. Vegetation is a significant factor in slope stability for three reasons: one vegetation adds weight to the slope, two its root system provide apparent cohesion to the slope material and third it provides a cover that cushions the impact of rain falling on slopes.

LAND MANAGEMENT

Land management embraces planning on rural, urban and regional scales. Rural land use management involves recognition and demarcation of land primarily capable of supporting forests, pastures and agricultural crops, in addition to dwelling abodes or structures. Forest land management is tantamount to management of the upland watersheds, and crop land management means analysis of the capability of soils and availability of water for irrigation and drinking.

Urban land use management encompasses habitat planning on the basis of conditions of stability of the ground for the foundation of structures, freedom from natural hazards, availability of water of requisite quality in sufficient quantity, suitability of sites for disposal of wastes, development of sites for recreation, roads, water supply etc.

Following are the four major elements of land management programme:

(i) Preparation of an inventory of land-resources, making use of aerial and satellite data complementary to comprehensive field work, and the delineation on maps.

(ii) Identification of natural hazards likely to threaten the area, indicating the anticipated degree and frequency of the threats.

(iii) Investigation of various geomorphic, geological, hydrological, pedological and ecological properties of the lands, and formulating a policy for land use commensurate with the needs of society.

(iv) Monitoring changes that occur consequent on the use of the land.

CHAPTER-VIII

ECOSYSTEM

CONCEPT OF AN ECOSYSTEM

The term ecosystem was first used by A. G. Tansley (1935) and defined it as an integrated system resulted from interactions of living and non-living factors of the environment. The concept of ecosystem is not so recent, Karl Mobius (1877) wrote about the community of organisms in an oyster reef as a biocoenosis. S.A. Forbes (1887) used the term microcosm for such an ecological system. According to Odum (1963), organisms and physical features of habitat form an ecological complex or ecosystem. Thus, ***ecosystem is the basic functional unit of ecology embracing biotic communities and abiotic environment both enfluencing each other.*** However, in a true sense an ecosystem may be defined as the ***basic structural and functional dynamic unit of ecology which harbors communities in its characteristic environment, interacting with each other and their environment independently as well as interdependently, preserving its identity in as far as possible in a self sustained manner.***

Thus an ecosystem has its own environment, where energy flow is sustained and maintained with the contribution of its structural constituents. It represents the highest level of ecological integration which is energy based.

An ecosystem may be as small as a pond, an agricultural field or as large as an ocean, desert or a forest. These unit ecosystems are simply separated from each other with time and space, functionally, they are all indeed linked with each other.

STRUCTURE AND FUNCTION OF ECOSYSTEM

Since an ecosystem is essentially a living and working dynamic system where energy requirements are obtained, created and dissipated, this simply means that there must be structural components in the system. Further since an ecosystem can better maintain its integrity if it bears the quality of self-sustenance, even then it must have all such components which would fulfill all its requirements from abiotic to biotic relationship. For that first we must have knowledge about the principle steps in the operation of an ecosystem. According to Clarke (1954), these principle steps are:

1. Reception of energy
2. Production of organic material for producers
3. Consumption of this material by consumers and its further elaboration
4. Decomposition of inorganic compounds and
5. Transformation of forms suitable for the nutrition of the producers.

Now if the area is inhabited by a self sufficient community, all these steps will be carried out smoothly and without any burden, otherwise materials or components must be incorporated from ambience. In brief all these five steps would cause, production, growth, death, decomposition of living material and their transformation to such a state whereby these can be reutilized and can enter life activities in inorganic form, thus this last activity would influence the non-living aspects of the habitat. That means, a lot of raw energy material would accumulate in the decomposed organic form, where inorganic elements and compounds become available in easily accessible form to producers and micro-organisms.

STRUCTURAL COMPONENTS OF AN ECOSYSTEM

The ecosystem has following structural components :

A. **Abiotic:** These include (a) Physical Factors: light, temperature, water, soil etc., and (b) Chemical factors: pH, various nutrients and gases.

B. **Biotic:** The biotic components are : (I)Producers and (b) Consumers, these may be, primary consumers (Herbivores), secondary consumers (Carnivores), tertiary consumers (Carnivores) and decomposers (Microorganisms or Micro-consumers)

A. The Abitoic (physico-chemical) components

The role of each physico-chemical factors is explicitly known to us vis-à-vis our ecological system. However, most important is the source of energy in the form of light. Light is the source of radiant energy, obtained from the sun, captured and utilized by the plants bearing green color pigment chlorophyll, to synthesize energy rich building organic materials, for themselves and for the consumers. Thus the sun occupies a pivotal place in our ecosystem, where radiant energy is converted into chemical energy by plants with the help of carbon dioxide, nutrients from soil and water or humidity as transportation medium for the nutrients. A self sustained ecosystem must have a ready and inexhaustible source of nutrient pool, which gets replenished without its efforts and open area to receive sufficient sun light to sustain its producer components, primarily. Some the important inorganic elements which are required by biotic components, almost essentially are N, K, Na, Cl, P, S, Co, Fe, Mg and Zn etc.

B. The Biotic (living) components

The living components belong to either of the two kingdoms viz. plants and the animals. Plants bearing chlorophyll are producers while all animals are basically consumers.

FUNCTIONS OF AN ECOSYSTEM

The study of ecosystem becomes meaningful to us only when we are in a position to know the productivity (or the production capability) of an ecosystem. The productivity of an ecosystem refers to the rate of production i. e. the organic matter synthesized and accumulated in any unit time. And the efficiency of any ecosystem depends much upon the rates of production of its primary producers. Our oceans are the largest ecosystem, though their productivity varies under different climatological, structural and regional levels. On the shore the productivity is around 2-3.5 gm m-2 day-1 , while at deep sea level its is only 0.5 gm m-2 day-1 . In a fertile lake or a pond it can be 5-10 gm m-2 day-1 or even more up to 30-50 gm m-2 day-1. On the other hand net productivity of crop plants may be between 0.25 kg to 12.0 kg m-2 year-1 . The productivity of any system can be enhanced with improvement of various eco-climatological conditions, like increase in day length i.e. light periods or better congenial temperature, better use and addition of fertilizers. Productivity is of the following types:

(i) Primary Productivity

Primary productivity is the activity of autotrophs, mainly by photosynthesis and also due to chemo synthetic microorganisms. All green coloring plants, phytoplankton and most bacteria fall in this category. It can be defined as –the rate at which radiant energy of sun is captured or stored by the producers (i. e. the photosynthetic and chemo synthetic activities). It can be classified under following two types:

(a) Gross primary productivity

It is the total rate of photosynthesis, which includes the organic matter used up during respiration at the time of estimation of the productivity. This is also some times referred to as total (Gross) photosynthesis or total assimilation. It depends on the chlorophyll contents. The rates of primary productive is estimated in terms of either chlorophyll contents as chl\gm dry wt\ unit area or photosynthetic number i. e. amount of carbon dioxide fixed \gm chl\ hour.

(b) Net primary productivity

This is the rate of production stored in the form of organic matter, in excess to the respiratory utilization of the productivity. In other terms , net primary production is the rate of increase of biomass and is also referred to as apparent photosynthesis or net assimilation. It is thus the balance between gross photosynthesis and respiration and other plant losses.

(ii) Secondary Productivity

It refers to the consumers or the heterotrophs. This is the rate of energy storage at consumer level. The consumers only utilize food materials, which has already been produced by the producers in their respiration, simply converting the food matter to different tissues by an overall process. It is not divided into gross or the net amounts. Secondary productivity is also called as rate of assimilation by ecologists like Odum (1971) and is in fact highly mobile and is not a fixed amount like primary productivity. The mobility means its assimilation rates differ from organism to organisms.

(iii) Net Productivity

Net productivity is the rate of storage of organic matter not used by the heterotrophs i. e. the consumers. It is usually equivalent to net primary production minus consumption by the heterotrophs during the unit time, like day, month, season or year. It is thus the rate of increase of biomass of the primary producers which has been not utilized by the consumers. It is generally expressed as production of cal gm m-2 day-1 .

Concepts of Productivity

According to Clarke (1954) the three fundamental concept of productivity are :

1. Standing crop
2. Material removed and
3. Production rate

1. The Standing Crop

It is the abundance of the organisms existing in the area at the time of observation, it may be expressed as number of individuals, as biomass, as energy contents or in some other suitable terms.

2. Material removed

The second concept of the productivity is the material removed from the area per unit time, which includes the yield to man, organisms removed from the ecosystem by emigration and material withdrawn as organic deposits. For a better self sustained ecosystem, material in some form must be supplied to it in an amount at least equal to the total amount of material removed from the area in the various ways, like harvesting or emigration and locking up of usable abiotic components in soil strata, which is not immediately available to the producers. However, the natural supply of materials in an area can be augmented artificially by adding fertilizers or by stocking.

3. Production Rate

The third concept of productivity is the production rate or the rate at which growth processes are going forward within the area.

Each of the fundamental concepts of productivity thus plays an important role. In brief the productivity is the sum total of activities of growth and production in an ecosystem, which includes the number of participants, their movement and the rate of their activities.

The fundamentals of productivity are more closely related with the primary producers, than to the other components. As primary production differs for different ecosystems, depending upon their producer components and the input or the availability of the abiotic variables. The size of producer has nothing to do with the rate or amount of production and productivity. The productivity in fact depends upon the rate of fixation of radiant energy in an ecosystem and not the size of the producers. The size of an ecosystem or biome also has nothing to do with the efficiency or rate of production. For example more than two third part of our planet earth is covered jointly by ocean and deserts, but terrestrial ecosystems of forests and grasslands or even a small lake in the tropical zone are more productive in terms of per unit area and unit time factor. Primary production in tropical forests ranges between 1000-3000 gm m-2 day-1 , while it is only about 600-2500 gm m-2 day-1 for the temperate forests. However, there area always variations in the productivity rates of different forests or grassland categories because many temperate forests are more productive than the tropical ones. For example on an average, highest rate of productivity has been found for the tropical rain forest ecosystems with an average of 2300 gm m-2 day-1 .

THE PRODUCERS

The producers are basically chlorophyll bearing plants which are capable of trapping light energy from sun, synthesize complex organic food material, with the help of carbon dioxide and water and release oxygen as a by product. More complex food required by these plants is also synthesized with the help of a variety of simple inorganic elements obtained by the plants from soil. Since all green plants are capable of synthesizing their own food material, hence these are also called as autotrophs (i.e. self nourishing). Thus in fact it is the beauty of the plant community only that it is capable of converting lifeless simple inorganic raw material into complex organic material or components of life forms, like carbohydrates, lipids and proteins. It is in these and various other complex organic substances that radiant energy of sun is fixed in the form of chemical energy at chemical bonds, which when break down with in living systems, liberate or release that trapped energy to be used by the consumers for their growth and development. Thus producers occupy the basic ground position in any ecosystem, over which the castle of ecosystem stands.

THE CONSUMERS

The consumers are all those living organisms, plants or animals which are not capable of synthesizing their own food material, from the simple raw material available in nature, hence they fully depend for their nourishment and energy requirements on the producers, i. e. the plant community. Therefore, these consumers are also termed as heterotrophs (nourished by others). Based on their feeding habits and preferences the heterotrophs or consumers are of two main types : Herbivores or the primary consumers and carnivores or secondary consumers.

(i) Primary consumers

These animals are purely herbivores and feed over producers i.e. the plants only, examples are rats, locust, butterflies, rabbit, cow and elephant etc.

(ii) Secondary consumers

These are animals which feed over the herbivore animals i.e. over the primary consumers and are thus carnivores or predators. These include frog, large fish, cat, snake, lion etc. The secondary consumer which eats a primary consumer is also called as predator and the primary consumer eaten by this predator is also called as prey. Thus there is a prey-predator relation ship among few animals, which plays important role in balancing the consumer population in any ecosystem.

(iii) Tertiary consumers

The tertiary consumers are all, again, carnivores and preferably predators, which prey over or feed over the secondary consumers, like python, kite, large sized fishes which eats medium or smaller sized fish and snakes.

There are many animals which eat whatever they get i. e. vegetarian food or carnivores diet, such organisms are called as omnivores (all eaters). Besides there are many organisms which strike or inhabit inside or over the body of other animals and derive their nutritional requirement from their hosts only and never synthesize their own food material, such organism are called as parasites. A parasite is an organism which usually does not kill its host at once, but takes its nourishments from host over a long period of time, till these are (host and parasite) in association. Besides there are organisms which derive their nourishments from the dead organisms., plants or animals, such organism are called as saprophytic or saprozoic.

(C) DECOMPOSERS AND TRANSFORMERS

These are often also called as micro-consumers. These are usually minute sized or microorganisms like protozoa, bacteria and fungi etc. Many insects are also decomposers and detritus feeders. The saprophytes are also among decomposers. All these derive their nutritional requirements from dead organisms only. Many organisms specifically feed over the decomposing organic material like rotten leaves, animal carcass etc., and convert these into some sort of digested or decomposed organic material, which is as good as ripe organic manure ready to be reused by plants as nutrients. Such organisms are detritus feeders and are some times called as transformers also, because these transform the decomposed organic matter fully, by breaking down complex substances into simple usable inorganic or organic salts and compounds, which can be taken up by producers.

Among metazoans the common earthworms play multiple role of scavenger, detritus feeder, decomposer as well as transformer. Such organisms are highly valuable for a sustainable ecology and environment, as these help in the conversion of complex organic material into raw nutrient material, which is added to the nutrient pool of soil. However, mostly the decomposers, detritus feeders and transformers have the similar mode of feeding and similar end product formation. Perhaps these can be regarded as the terminal link between consumers and the abiotic substances.

ENERGY FLOW IN AN ECOSYSTEM

Energy is the capacity to do work. Biological activity requires utilization of energy, which ultimately comes from the sun. Solar energy is transformed into chemical energy by the process of photosynthesis, this is stored in plant tissues and then transformed into mechanical and heat forms during metabolic activities. In the biological world, the energy flows from the sun to plants and then to all heterotrophic organisms.

Mechanical energy has two forms, namely kinetic energy or free energy and potential energy. The energy a body possesses by virtue of its motion is called kinetic energy and is measured by the amount of work done in bringing the body to rest. This includes animal activity as well as heat or thermal energy and light. Potential energy is stored energy and useful after conversion into kinetic energy. All organisms require a source of potential energy, which is found in the chemical energy of food. The oxidation of food releases energy which is used to do work. Thus, chemical energy is converted into mechanical energy. Energy can be transformed from one form to another e. g., the potential energy in the chemical bonds of a molecule of glucose can be converted to heat energy or movement. Light energy can ultimately be converted to chemical energy when it is absorbed by specific molecules in plants. Energy transformation is governed by the laws of thermodynamics. Of particular relevance is the observation that different forms of energy can be arranged along a quantitative scale of disorder, called entropy (it is a measure of energy not available due to transformation of energy at each successive trophic level in the food chain), from highly organized energy, random energy, and that every process or transformation involving energy leads to an increase in entropy. Light and chemical energy are highly ordered forms of energy that can be used by organisms. In living systems, most energy ultimately becomes transformed to heat, which is a random form of energy that cannot be further transformed by organisms.

“G = “ H – T “ S, G= change in the free energy of the system; H= change in enthalpy, which is a change in the amount of energy in the form of heat liberated or absorbed by the system during physical or chemical changes; S= entropy change of the system and T= absolute temperature

The most commonly used measurement units of energy is Joule. One Joule is the energy required to raise one kilogram of mass to a height of 10 cm in earth's gravity. One Joule is equal to 0.239 calories. The calorie is a measure of energy that only applies to heat.

Animals obtain chemical energy by eating other animals or plants. Much of the energy contained in consumed food is passed to the chemical bonds of a molecule called ATP by way of an oxidation process called respiration. The potential energy stored in ATP can later be made available to the cells to perform work.

Laws Governing Energy Transformation: Energy transformations are governed by the laws of thermodynamics which are usually, applied to closed system. The first law of thermodynamics is the law of conservation of energy, which says that ***energy may be transformed from one form into another but is neither created nor destroyed.*** If an increase or decrease occurs in the internal energy (E) of the system itself, work (W) is done and heat (Q) is either evolved or absorbed. Thus

$$E = W + Q$$

Where E is the decrease in internal energy of the system, W is the work done by the system and Q is the heat given off by the system.

The total amount of heat produced or absorbed in a chemical reaction, either occurring directly or in stages, always remains the same. This is called the specific law of constant heat, and is included in the first law. This law recognizes the inter convertibility of all forms of energy but does not refer to the efficiency of transformation or conversion. In ecological system, solar energy is converted into mechanical and heat energy. Hence energy is not created or destroyed in ecological system but is converted from one form to another.

The second law of thermodynamics states that ***processes involving energy transformation will not occur spontaneously unless there is degradation of energy from a non-random to random form.***

In man-made machines (closed system), heat is the simplest and most familiar medium of energy transfer. But in biological system its is not a useful medium of energy transfer, as living systems are essentially isothermal and there are no significant difference in temperature between different parts of cell or between different cells in tissue. Thus cells are not heat engines.

The Functional interrelation between living organisms and their habitat constitute an ecosystem. Energy is the driving force for an ecosystem as each and every activity of an organism requires energy, and the output of such activities depends on the energy available to and mobilized by the organism. The ultimate source of energy for ecosystem or for the biosphere as a whole is the sun. The flow of energy ecosystem along with the pattern and efficiency of energy utilization is an important ecological process and comprises the field of ecological energetics. Green or photosynthetic plants are first to absorb solar energy in an ecosystem and transfer it to animals in the form of food, hence there is wide variation in energy flow patterns within different ecosystems depending upon the availability of energy, the primary production pattern, the energy capturing efficiency of plants, energy losses at different steps of energy transfer, ecosystem structure and various interrelation between and within biotic and abiotic components.

Capturing energy from solar radiations and fixing it in the synthesized organic material, is the first step of energy conversion from purely abiotic components to biotic (organic) constituents of ecosystem. The flow of energy is thus essentially from sun to producers and from producers to consumers of various categories and is almost a unidirectional phenomenon in nature. This leads us to show that once energy has been trapped by the producers it can not be reverted back to the sun or abioitic ingredients which participated in fixing this energy among the producers. It is thus that the solar energy (radiation energy) is converted into chemical energy. From producers, energy is transferred to higher trophic level of primary-herbivore consumers, in the form of their food, which is again transferred by the primary consumers to secondary or tertiary consumers by way of assimilation efficiencies at successive trophic levels, in the form of food material only. A lot of energy is consumed by the organisms at various trophic levels during growth, repair, reproduction, respiration, excretion etc. The balance of the energy present in the biomass at any trophic level gives the Gross Net Production. It is therefore, that lesser energy is available at successive higher trophic levels, even if the biomass is more.

A single channel energy flow model has been proposed by Lindaman (1942). It shows that out of about 1,18,872 gm cal cm-2 yr-1 of available solar energy, only about 111 gm cal cm-2 yr-1 is utilized by the autotrophs. This means just 0.10% of solar energy is being utilized by the producers of our ecosystem, while 99.9% remains unutilized. Out of the total energy being utilized, about 21% or 23 gm cal cm-2 yr-1 is consumed in metabolic reactions of autotrophs for growth etc., 15 gm cal cm-2yr-1 is consumed by herbivores, which feed upon the producers (or the autotrophs) this is equal to 17% of the net autotrophs production. 3.4% of the net production goes for the decomposition. 79.5% of net production is not utilized at all, but becomes part of the accumulating sediments. Thus collectively three major fates i. e. decomposition, herbivory and non utilization, are equivalent to net production. Out of about 15 gm cal cm-2 yr-1 available at herbivores level, about 30% is used in metabolism. Thus more energy (30%) is lost by herbivores, than by the autotrophs (21%). Again there is considerable amount of energy available for the carnivores, which is 10.5 gm cal cm-2 yr-1 or 70%, again is not fully utilized, and only 3.0 gm cal cm-2 yr-1 or about 28.6% of net production passes to the carnivores. At carnivore level about 60% of the carnivore energy intake is consumed in metabolic activity, while the remainder becomes part of the unutilized sediments, thus only very little part of this is subjected to decomposition, yearly. This high respiratory loss compares with 30% by herbivores and 21% by autotrops in this ecosystem.

ECOSYSTEM DYNAMICS

The nature is highly diversified in its make up. Our earth is most unique in our solar system which holds three distinct types of structural moieties viz. (a) Land – the lithosphere- with soil and its solid constituents, (b) Water – the hydrosphere – with liquid components and, (c) Air – the atmosphere – with gaseous components. However, all strata of these basic components of our earth are not habitable and only a comparatively thin area in and around earth is habitable for the living organisms of all varieties. The parts of this planet earth where water, air and soil come in contact with each other and interact, provides most characteristic zone, which is capable of habitation of life forms in its myriad diversity. This zone of nature of earth where life can sustain is called as Biosphere. Under natural conditions all living entities are co-hebetating, interacting and interdependent in such a manner that a stable sustainability of equilibrium is maintained. Any small alteration in any one of its basic components could lead to collapse of this natural grand ecosphere of this planet earth. The life is , thus found only up to a limited depth, height and specific areas of biosphere of geological, geographical, climatological, contributions, hence the life support systems of our ecosphere are also equally diversified and distributed all over this globe. As a result different types of biospheroic zones have developed all over the earth, sustaining to characteristic types of aquatic, marine, mountain, grassland and other types of biospheric realms. In brief the whole of our earths biosphere is characteristically made up of the different types of habitations. Each type of habitation is very unique in itself and forms a specialized ecological system of its own and tries to sustain its biological constituents. Thus each of such distinguished habitat with its own characteristic flora, fauna and environmental realm or area with strong tendency of self-sustenance forms a specific type of ecosystem. Since we have primarily two basic types of established habitats viz. Aquatic and Terrestrial, accordingly we have ecosystems differentiated in these two habitat realms only. These ecosystems are divided into various types. Very large terrestrial ecosystems are called as Biomes. A biome may be defined as the largest terrestrial ecological unit characterized by interactions of flora, fauna and various other abiotic substances. Thus the biomes are earth's major ecosystems. The earth has been divided into following seven major ecosystems or biomes which are named after their climax or dominant plant form:

1. Temperate forest biome
2. Grass land biome
3. Desert biome
4. Coniferous forest biome
5. Tundra biome
6. Savanna biome and
7. Tropical rain forest biome

Similarly the aquatic ecosystems can be divided into following types :

(a) Fresh water ecosystem

This in turn can further be dived as, pond ecosystem, river ecosystem and lake ecosystem.

(b) Marine ecosystem

The marine ecosystem is by and large more uniform than the terrestrial and the fresh water ecosystem and can be divided into : Estuarine ecosystem, Oceanic ecosystem and Deep sea ecosystem

These divisions of major ecosystems can be further classified into sub units. As each major ecosystem or the biome in itself contains a number of smaller units.

The division of the landscapes into different ecosystems is rather arbitrary. These seldom have distinct boundaries, and often one type of ecosystem or biome stretches or encroaches into other type of biome. Thus we have blends of two types of biomes in certain areas. These areas are called as transition zone or ecotones. These transition zones are unique areas of an ecosystem often harboring more diversified flora and fauna with higher bio-diversity than in the contributing ecosystems on either sides, and it has been found now that these blending zones or transitional zones are better areas of bio-diversity and ecological evolutions, due to higher variations of life forms found therein. Further no eco-system is isolated or works in isolation. Despite very unique and characteristic types of climatological and environmental conditions, each ecosystem is continuously under influence of biotic as well as abiotic components of other ecosystems. For example a flood can contribute a lot of new soil to grassland or a river or a pond or certain fauna can visit another major ecosystem during a particular season of the year or for a particular reason or purpose. However, by and large each type of ecosystem tries to maintain itself in a self sustained manner, so that its integrity and uniqueness is not disturbed or is rather preserved.

As mentioned above, an ecosystem is a dynamic system, where all living and non living components of our nature are interacting with each other in a very balanced and harmonious manner, so as to keep it as self sufficient and sustainable as possible to preserve its identity. Hence, any part of land or water body within an area consisting of specific flora, fauna and physical conditions, like a desert, mountain, forest, an orchard, crop land or a river basin, stream, a lake or a pond, a bay can be taken up to study and understand the structure and functioning of a balanced ecosystem. A brief description of the four major types of ecosystems is given below:

THE FOOD CHAINS

It is quite evident that in any natural ecosystem, there exists a definite cycle between transformation of abiotic components to biotic components and vice versa, as well as there is recycling of energy and flow of energy from one level (abiotic to biotic) to another level. The producers capture the radiant energy of sun, photosynthetically bind it in various organic compounds, use it themselves, also provide (transfer) their bonded energy to heterotrophes, where energy moves from one level of consumer to another higher level of consumers, viz. from herbivores to primary consumers, to secondary consumers to tertiary consumers. In nature generally we find two types of food chains :

1. Grazing Food Chain

This is the most characteristic type of food chain. It starts with the producers i. e. plants and the phytoplankton, grazed or eaten up by the herbivores, which in turn are eaten by the secondary consumers or the carnivores and the predators. This type of food chain in an ecosystem is most prevalent type, which directly depends on the availability and utility of radiant energy of the sun. Thus autotrophs play prominent role here. Following figures show a food chain in terrestrial and in an aquatic ecosystem.

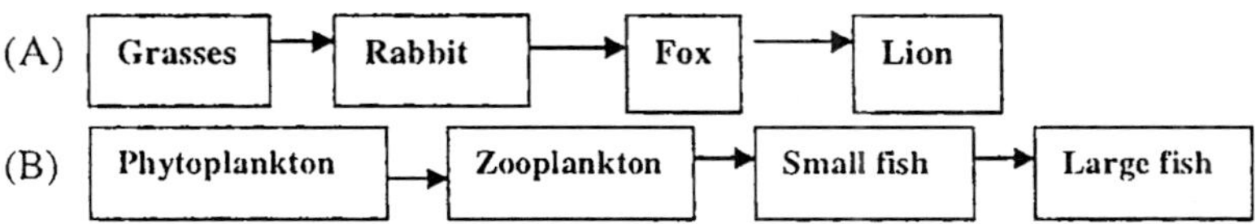

2. Detritus Food Chain

This type of food chain begins with the dead and decomposed organic matter in the nature. Microorganisms feed over the dead and decaying organism (detritivores) of all categories i. e. over the producers and the consumers as well. Thus this food chain is not directly dependent on solar energy, but on the energy derivatives of another system. However, this type of food chain is usually important in nature as it releases the complex organic energy into reusable ingredients of organic or inorganic compounds or salts, thus contributing to the nutrients pool of the nature or any ecosystem. The scavengers like crabs, saprobes, saprozoic organisms and even earth worms also some time fall in this category.

But ultimately both types of food chains merge with one another or are linked to each other, as when all these biotic components die, perish, decay and are decomposed and transformed to reusable form by decomposers, while the energy contained is dissipated, used and released into their ambience. Thus it leads us from a simple concept of food chain to a more higher grade of organized food web concept operating in nature in all ecosystems.

THE FOOD WEB

It is apparent that no food chain in our nature can work in isolation. That is the living components of nature by way of their feeding habits are equipped with alternatives and substitutes. Each food chain has components which also form the food component for the animals of other food chain. Hence various food chains operating in an ecosystem are interconnected at various places with other food chains, thus an interlocking pattern is developed. These interlocking patterns of different food chains operating among living organisms in an ecosystem is called as the food web.

It amply makes it clear that no food chain system of energy flow from one organism to other or the energy transfer from one trophic level to other trophic level move in a linear or straight line manner. But that longer the food chain greater are the opportunities of interlinking and interlocking of components with each other. This can be illustrated from the following example :

1. Phytoplankton → Zooplankton→ Rotifer → Fish → Kingfisher (Bird)
2. Phytoplankton→ Protozoan→Rotifer→ Crustacian→Fish
3. Phytoplankton-----Rotifer→ Mollusc-→ Small Fish→ Large Fish →Man
4. Cereals→Man
5. Grass →Grasshopper → Lizard →Hawk
6. Grass → Rabbit→ Dog
7. Grass→----Insect→-----Frog-→------Snake
8. Grass→-----Rat→-----Snake→-------Hawk

9. Tree→---Insects→-----Birds-→----Snake→—Hawks
10. Tree→----Birds→-----Parasites→-----Snakes→—(Dead) →----Decomposers
11. Decomposed leaves→----Earthworms→-----Birds
12. Decomposed organic matter→---Saprobes→—Zooplanktons

All the above given examples clearly show that no food chain is isolated one. And that shorter the food chain lesser would be interlinks or interceptions and longer the food chain more will be interconnections and side food chains. And when ultimately all living organism are dead, the decomposers and transformers convert these into simple organic or inorganic compounds, which go to the nutrient pool of the system. Thus we find that energy flows from abitoc level to biotic producers and consumers is in a stepwise manner. These various stages and steps of energy transfer are called as trophic levels, where energy moves (is transferred) from one level to another. This can be very well illustrated with the help of certain ecological pyramids.

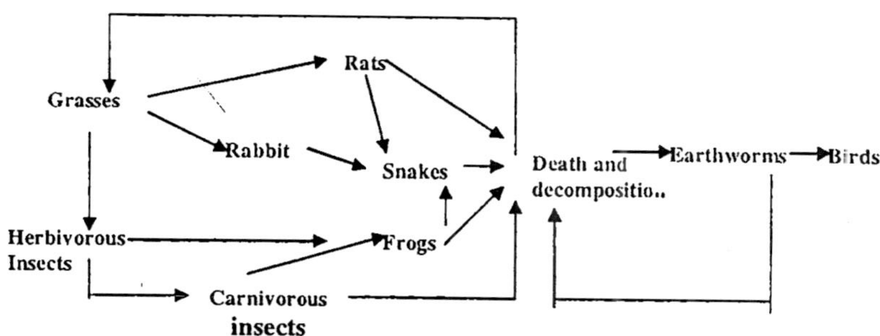

A simple food web

THE ECOLOGICAL PYRAMIDS

It is evident from fore going details that there is always some sort of relationship among the biotic components in any ecosystem. This relationship influences each other and the abiotic components as well. This relationship can be easily observed in the Number, Biomass and Status of Energy Flow in any ecosystem. Diagrammatically or graphically these relationships can be demonstrated with the help of pyramids. In any ecosystem it is first the number i.e. population of its community components, then their size showing their biomass and this biomass shows energy status at successive trophic level.

1. Pyramid of Numbers

As we know that in any ecosystem all heterotrophs depend for their energy requirements on the producers, hence the population dynamics in terms of number of organisms at different trophic levels varies in accordance to the numbers and size of the producers and consumers. The pyramid of number thus shows a relationship between the numbers of producers and the consumers. The base naturally shows the number of primary producers, then the successive steps of the pyramid reveal the number of primary consumer, followed by the number of the secondary and then the tertiary consumers, thus the top of the pyramid represents the top carnivore in the ecosystem.

Usually these pyramids are up right, for example in an grassland, grasses at the base forming the producers level, are maximum in number, then at secondary trohic level the rats and rabbits or grasshoppers are lesser in number, then at third trophic level the number of secondary consumers, which are always carnivores like frogs, foxes, birds is lesser and then at fourth level the numbers of tertiary consumers like a snake, vultures, lions etc. is still lesser and in lowest in this pyramid of numbers. Thus we find that the number of organisms at successive level goes on decreasing as we move from base to the top level of pyramid of numbers.

On the other hand a pyramid of number can be inverted. If we take the example of a tree as an ecosystem of its own in a forest. The tree is a producer, inhabited by a large number of herbivore birds, which in turn may be harbored with still greater population of parasites. Thus here producer is just one tree, primary consumers are several birds followed by still greater population of parasites at tertiary level, making an inverted pyramid.

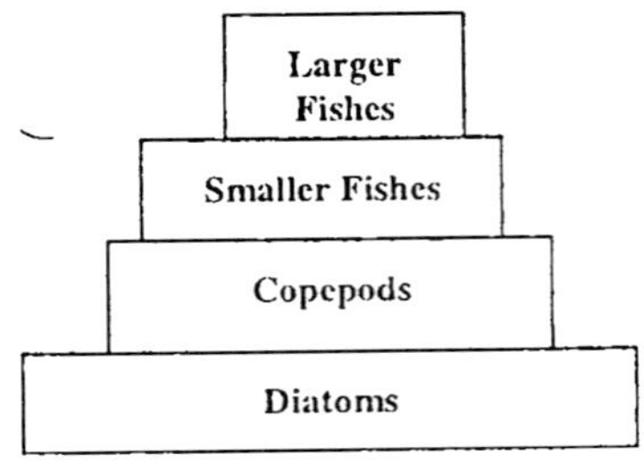

Pyramid of numbers in a pond ecosystem

2. Pyramid of Biomass

The pyramid of biomass shows the relationship of population supported at each successive trophic level in terms of real mass of the standing crop, that is the quantitative relationship of the standing crop. In a grassland and forest ecosystem, there is generally a gradual decrease in biomass of organisms at successive levels from the producers to the top carnivores. As a result the pyramid for these ecosystems are usually upright. But in a pond or river due to small size of producers, which are mainly phytoplantkons, their biomass is minimum at the base, and it goes on increasing towards apex level of the pyramid, where top carnivore is present, thus the pyramid is an inverted one.

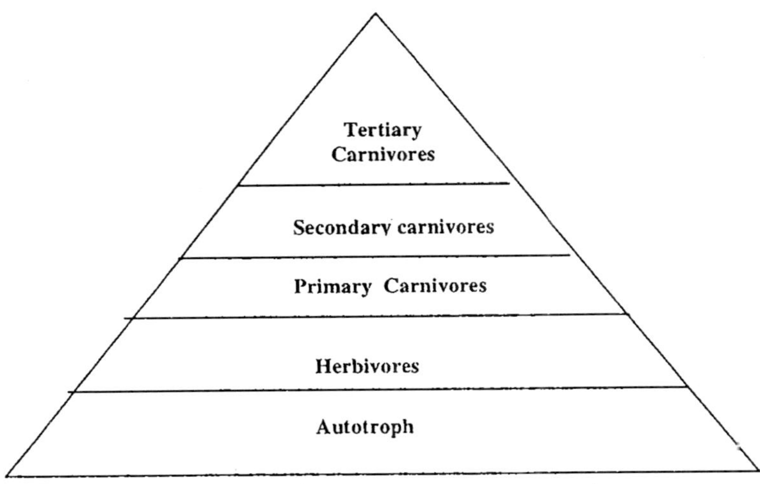

Pyramid of biomass in a terristrial ecosystem.

3. Pyramid of Energy

This is the best way to express the overall nature of an ecosystem. Here the pyramid structure is based not on the amount of energy fixed at successive levels from producers to the top consumers, but rather on the rate at which food is being produced. Thus as against the number and total biomass of components at successive levels, in the pyramid of energy it is the rate and amount of energy in real terms available at successive levels, as it passes from base of producers to the apex level of carnivores along the food chain. It is always an upright type of structure. This also helps us to understand that how much energy is lost in transfer of energy from one trophic level to another, due to dissipation of heat energy or in the process of growth and development of the components of higher trophic levels, and only how much energy is in fact available there, as a result of productivity. As Clarke (1954) has rightly pointed out that at each step in the food chain, loss of energy and of material from the system takes place because the process of assimilation and growth are not 100 percent efficient. This means that the organic matter produced per average unit of time and the energy represented by it, becomes less at each successive trophic level.

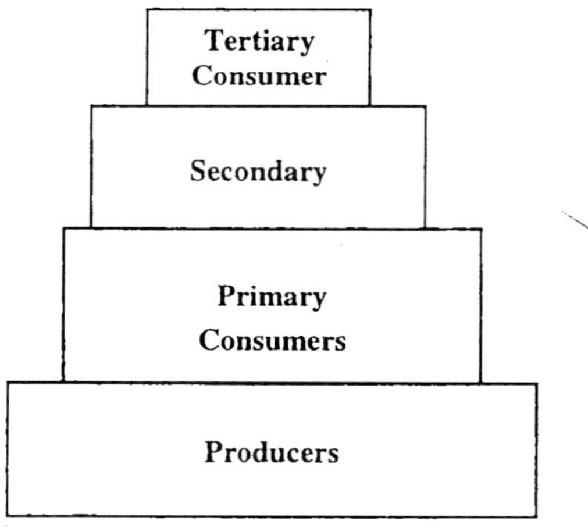

Pyramid of energy

DIFFERENT TYPES OF ECOSYSTEMS ANDTHEIR CHARACTERISTIC FEATURES

FOREST ECOSYSTEM

Forests are large terristrial habitats and constitute one of the important ecosystems of the world. The different components of a forest ecosystem are as under:

The abiotic components include various organic and inorganic substances besides the dead organic material found in the forest soil. The availability of light in a forest ecosystem may vary depending upon the types of species and their canopy structure.

The different biotic components of a forest ecosystem iare as under:

(i) Producers

Trees are the main producers in a forest ecosystem which have a high species diversity and show a great degree of stratification in tropical moist deciduous forests. A forest also has a good diversity of shrubs and ground vegetation. In a tropical deciduous forest the important producers include Shorea robusta, Mallotus philipinensis, Tectona grandis etc.

(ii) Consumers

The different consumers in a forest ecosystem are as below;

(a) Primary consumers

The herbivorous animals which feed on different plant parts constitute the primary consumers and include butterflies, grasshoppers, bugs and leaf hoppers, various grazing animals such as rabbits, squirrels, mongooses, deer and the largest land animal the elephants.

(b) Secondary consumers

These are the animals which feed mainly on primary consumers and include lizards, snakes, birds and fox etc.

(c) Tertiary consumers

Carnivores like tigers, lions, hawks etc are the tertiary consumers.

(iii) Decomposers

Micro-organisms including bacteria, fungi etc. which convert the dead parts of plants and trees in tho organic material and finally in to the smaller elements constitute the decomposers.

GRASSLAND ECOSYSTEM

These are again a very important types of ecosystem and are known as prairies in North America, steppes in the Eurasia, pampas in Argentina, veldt in South Africa. The grassland constitute 20 % towards the total land surface and may be either tropical, temperate or alpine. The different components of a grassland ecosystem are as under:

1. Abiotic Components

The various types of nutrients present in the soil e.g., C, N, K, P, S etc. constitute the abiotic components of a grassland ecosystem.

2. Biotic Components

(i) Producers- The producers in a grassland ecosystem are mainly the different types of grasses and forbs and the species like Cynodon dactylon, Desmodium, Setaria, Digitaria etc are very common.

(ii) Consumers: The herbivores feeding on different types of grasses are the primary consumers and include rabbit, mouse, sheep and some other invertebrates.

The secondary consumers which feed on herbivores include jackals, frogs, snakes some birds etc.

3. Decomposers

The various species of microorganisms which are very active in decomposition of dead organic matter are bacteria, fungi and some actinomycetes.

DESERT ECOSYSTEM

The desert ecosystems are either do not have any vegetation or it is very scanty that too in the form of thorny bushes. These occupy about 17% of the terristrial ecosystems and receive an annual rainfall of less than 25cm which is confined to only a maximum of 50 days of a year. These are classified as warm deserts e.g., deserts of Sahara in northern Africa, Kalahari in Southern Africa and Thar in India and cold deserts e.g., deserts of Iran and Turkey, the Gobi desert of Mangolia. Some species of shrubs, bushes, some grasses are the principal producers of a desert ecosystem. Some sacculent species like cacti and xerophytic mosses are also present.

Consumers of a desret ecosystem include reptiles, some nocturnal rodents, birds and insects which are able to live in such dry conditions. Decomposers are not many and include some fungi and bacteria most of which are thermophilic.

AQUATIC (A POND) ECOSYSTEM

A pond is usually a small water body, not deep enough to show any thermal stratification. Under normal ecological conditions a permanent pond has its regular sources of water and is able to use the natural abiotic components like light, air, mixed nutrients from soil, heat of sun rays, water currents etc. and has its own characteristic biotic community, which by and large derives all its energy, growth, development related requirements from itself. It seldom depends for its physico-chemical and biological requirements on any other neighboring ecosystem. In brief we can summarize various components of a pond ecosystem as following:

1. Abiotic components

(a) The soil profile acts as a substratum, besides a source of nutrient elements, holds water and provides base for the growth of rooted plants.

(b) Water in a pond is the most essential component. It also provides a substratum and medium to aquatic biota for all related requirements.

(c) Sun light and radiant energy: These are available to a pond as a natural phenomenon, and influence the productivity and life rhythms of its living organisms, beside also affecting the interactions among the abiotic components-almost continuously. For example rate of photosynthesis, oxygen solubility in water, rate of respiration among biotic components, reproductive cycles among living organisms, salt and compound synthesis, nutrient cycling etc.

(d) Atmospheric gases are also available to a pond ecosystem in routine natural manner.

2. Biotic components

(i) The producers are exclusively the plant communities, which by the help of radiant energy of sun and other abiotic essentials photsynthetically act as producers.

The producers of aquatic systems include minute one celled algae to multi-cellular plants which may be floating, submerged or planktonic. The life cycle of these producers follows the natural course, dies, decays, decomposed and transformed into reusable micro-nutrients.

(ii) The consumers: Primary consumers include zooplanktons, protozoan, rotifers, coelenterates, herbivore insects, crustaceans, fish fry etc. The secondary consumers include predatory insects, carnivore fishes, frog etc. The tertiary consumers include large sized carnivore fishes, predatory birds like king fisher etc.

(iii) The decomposers and transformers are a variety of microorganisms, like protozoan, bacteria and fungi, which thrive over the dead organisms, as saprobes, besides many metazoan insects, mollusks etc. which also feed over the dead organisms and convert it into producer usable refuge.

CHAPTER-IX

ECOLOGICAL SUCCESSION

Communities are made up of populations, which interact in many ways and influence their development over time. It is a fact that in given biotopes certain communities tend to succeed one another. Since community is a dynamic system, and as such is never found permanently in complete balance with their component species or with the physical environment which keeps on changing over a period of time due to (i) variations in climatic and physiographic factors and (ii) the activities of the species of the communities themselves. These influences bring about marked changes in the dominants of the existing community, which is sooner or later replaced by another community at the same place. This process continues and successive communities develop one after another over the same area, until the terminal final community again becomes more or less stable for a period of time. This occurrence of relatively definite sequence of communities over a period of time in the same area is known as ecological succession.

Ecological succession, also known as 'Ecosystem Development' may thus be defined in terms of the following parameters :

1. It is an orderly process of community development that involves changes in species structure and community process with time.

2. It is reasonably directional and therefore, predictable.

3. It results from modification of the physical environment by the community i. e. succession is community controlled, though the physical environment determines the pattern, the rate of change and often sets limit as to how far development can go.

4. It culminates in a stabilized ecosystem in which maximum biomass and symbiotic function between organisms are maintained per unit of available energy flow.

The whole sequence of communities that replaces one another in a given area is called the sere, the relatively transitory communities are called serial stages or development stages or pioneer stages, while the terminal stabilized system is known as the climax. Species replacement in the sere occurs because population tends to modify the physical environment, making conditions favorable for other population until an equilibrium between biotic and abiotic constituents is established.

The strategy of succession as a short term process is basically the same as the strategy of long term evolutionary development of the biosphere, namely increased control of or homeostasis with the physical environment in the sense of achieving maximum protection from its perturbations. The development of ecosystem has many parallels in the development biology of organisms, and also in the development of human society. The extent to which ecological succession is self induced- as distinct from being caused by changes imposed, varies greatly in different situations. Similarly the predictability of the course and speed of succession is variable.

TYPES OF SUCCESSION

Ecological succession can primarily be separated into two components physio-graphic succession, influenced by all the physical features of the environment and biotic succession, determined by the animals and plants comprising the community itself.

1. Physio-graphic Succession

This form of succession was described by H. C. Cowles (1901) with respect to the series of communities that gradually arose as port glacial lake of Chicago drained away leaving the smaller lake Michigan and Chicago plain in its place. Cowles (1991) explained it that , having related the vegetation largely to topography, we must recognize that topography changes do not take place in an haphazard manner, but according to well defined laws. The process of erosion ultimately causes the wearing down of the hills and the filling up of the hollows. These two processes, denudation and deposition, working in harmony produce plantation and the inequalities are brought down to the base level. The chief agent in all these activities is water and no fact is better established than the gradual eating back of the rivers into the land and wearing away of coast lines, the material thus gathered fills up the lakes, forms the alluvium of flood plains or is taken to the sea. Vegetation plays a part in all these processes, the peat deposits adding greatly to the rapidity with which lakes and swamps are filled, and on the other hand the plant covering of the hills greatly retards the erosive processes. Thus the hollows are filled more rapidly than the hills are worn away. As a consequence of all these changes, the slopes and soils must change. This consequently brings about changes in the plant communities, which are replaced in turn by others that are adapted to the new conditions.

Successional changes of this kind are widespread and basic. Erosion and deposition of soil by both wind and water profoundly modify the earth's crust within a given climatic period. These changes alter the structure of terrestrial and fresh water communities to such an extent that established populations of plants and animals are unable to cope with the changed conditions and their place is taken by other species that can adjust, hence communities succeed each other through time.

2. Biotic Succession

The second component in ecological succession signifies change in community structure brought about through biological action of the plants and animals constituting the community. These processes and their effects collectively have been termed community development. For example waste products, feaces, decay of dead organisms, formation of organic soils, filing up of lake and pond bottoms with organic sediments and other influences gradually change a community and thus residents are unable to tolerate the changed conditions and are replaced by different species populations. Components of biotic succession can be further grouped as :

(a) Primary Succession

If development begins on an area that has not been previously occupied by a community (such as newly exposed rock or sand surface of a lava flow), the process is known as primary succession. The first group of organisms that establish themselves there are known as the pioneers, primary community or primary colonizer.

(b) Secondary Succession

If community development is proceeding on sites previously occupied by well developed communities (such as abandoned crop field or cut over forest which may be rich in nutrients, and survival conditions are favorable), the process is appropriately called secondary succession. Secondary succession is usually more rapid because some organisms or their disseminates are already present and previously occupied territory is more receptive to community development than sterile areas.

(c) Autogenic Succession

After succession has begun in most of the cases, it is the community itself which, as a result of its reactions with the environment modifies its own environment and thus causing its own replacement by new communities. This course of succession is known as autogenic succession.

(d) Allogenic Succession

In some cases, however, the replacement of the existing community is caused largely by any other external condition and not by the existing organisms. Such a course is referred to as allogenic succession.

On the basis of successive changes in nutritional and energy contents, succession are sometime classified as :

(i) Autotrophic Succession

It is characterized by early and continued dominance of autotrophic organisms like green plants. It begins in a predominantly inorganic environment and the energy flow is maintained indefinitely. There is gradual increase in the organic matter content supported by energy flow.

(ii) Heterotrophic Succession

It is characterized by early dominance of heterotrophs, such as bacteria, actinomycetes, fungi and animals. It begins in a predominantly organic environment and there is a progressive decline in the energy content.

In ecological literature, many other kinds of successions are mentioned, depending mainly upon the nature of the environment (primarily based upon moisture relations) where the process has begun, and thus it may be a Hydrosere or Hydrarch strarting in regions where water is in plenty, as ponds, lakes, stream, swamps, bogs, etc, a Mesarch, where adequate moisture conditions are present, and a Xerosere or Xerarch where moisture is present in minimal amounts, such as dry deserts, rocks etc. Sometimes these are further distinguished as, the Lithosere initiating on rocks, Psammosere on sand and Halosere in saline water or soil.

CAUSES AND PROCESS IN SUCCESSION

Since succession is a process, more properly a series of complex processes, it is natural that there may not be a single causes for this. Generally, there are three types of causes:

(i) Initial or Initiating Causes

These are climatic as well as biotic. The former includes factors, such as erosion and deposits, wind, fire etc, caused by lightning or volcanic activity and the latter includes the various activities of organisms. These causes produce the bare areas or destroy the existing population in an area.

(ii) Ecesis or Continuing Causes

In order for new species to invade an area they not only must have some means of reaching the new localities but also must be able to grow and reproduce under the conditions found there. Ordinarily the first invaders of a bare area are plants, and for them the physical features of soil and the climate are of primary importance in determining whether or not ecesis, or successful establishment , can take place. The success with which a species can extend into a variety of new regions depends on its ability to tolerate widely different ecological influences both, physical and biological.

(iii) Stabilizing Causes

These cause the stabilization of the community. According to Clement, climate of the area is the chief cause of stabilization, other factors are of secondary value.

The whole process of primary autotrophic succession is actually completed through a number of sequential steps, which follow one another. These steps in sequence are as follows :

1. Nudation

This is the development of a bare area without any form of life. The area may develop due to several causes such as landslides, erosion, deposition, or other catastrophic agency. The cause of nudation may be :

(a) Topgraphic

Due to soil erosion by gravity, water or wind, the existing community may disappear. Other causes may be deposition of sand etc, landslide, volcanic activity and other factors.

(b) Climate

Glaciers, dry periods, hail and storm, frost, fire etc. may also destroy the community.

(c) Biotic

Man is one of the most important biotic factor, responsible for destruction of forests and grasslands for industry, agriculture, housing etc. Other factors area disease, epidemics due to fungi, bacteria, viruses which destroy the whole population.

2. Invasion

This is the successful establishment of a species in a bare area. The species actually reach this new site from any other area. This whole process is completed in following three successive stages:

(a) Migration (dispersal)

The seeds and spores of the species reach the bare area. This process, known as migration, is generally brought by air, water, etc.

(b) Ecesis (establishment)

After reaching to new area, the process of successful establishment of the species as a results of adjustment with the conditions prevailing there, is known as ecesis. In plants, after dispersal, seeds or propagules germinate, seedlings grow, and adults start to reproduce. Only a few of them are capable of doing this under primitive harsh conditions, while most of them disappear. Thus as a result of ecesis, the individuals of species become established in the area.

(c) Aggregation

After ecesis, as a result of reproduction, the individuals of the species increase in number and they come close to each other. This process is known as aggregation.

3. Competition and Co-action

After aggregation of a large number of individuals of species at a limited place, there develops competition (inter as well as intra specific) mainly for space and nutrition. Individuals of a species affect each others life in various ways and this is called co-action. The species, if unable to compete with other species, if present, would be discarded. To withstand competition, reproductive capacity, wide ecological amplitude etc. are of much help to the species.

4. Reaction

This is the most important stage in succession. The mechanism of the modification of the environment through the influence of living organisms on it is known as reaction. As a result of reactions, changes take place in soil, water, light conditions temperature etc. of the environment. Due to all these, the environment is modified, becoming more unsuitable for the existing community which sooner or later is replaced by another community. The whole sequence of communities that replaces one another in the given area is called a sere and various communities constituting the sere, as serial stage, serial communities or developmental stages. The pioneers are likely to have low nutrient requirements, more dynamic and able to take minerals in comparatively more complex forms. They are small sized and make less demand from environment.

5. Stabilization (climax)

Finally there occurs a stage in the process, when the final terminal community becomes more or less stabilized for a longer period of time and it can maintain itself in equilibrium with the climate of the area. This final community is not replaced and is known as climax community and the stage as climax stage.

Some ecologist (Gleason 1929) have talked of retrogressive succession in which continuous biotic influences have some degenerating influence on the process. Due to destructive effects of organisms, sometimes the development of disturbed community does not occur and the process of succession, instead of becoming progressive, becomes retrogressive. As for e.g. forest may change to shrubby or grassland community. This is called retrogressive succession.

CLASSIFICATION OF SERES

As mentioned earlier, the seres are classified primarily upon the water content of the initial area in which they develop. The two important seres are: Hydrosere and Xerosere.

(a) Hydrosere or Hydrarch

Succession beginning in ponds, pools, lakes, marshes and else where in water are termed hydrarch and the different stages of series or sere constitute a hydrosere. Hydrosere, originating in a pond, starts with the colonization of some phytoplankton which form the pioneer plant community and finally terminates into a forest, which is a climax community together with their chief components of vegetation. In the initial stage, phytoplankton are the pioneer colonizer. They are consumed by zooplankton and fish. Gradually these organisms die and increase the content of dead organic matter in the pond. This is utilized by bacteria and fungi and minerals are released after the decomposition. The nutrients rich mud then supports the growth of rooted hydrophytes in the shallow water zone. The hydrophytes die and are decomposed by microorganisms thus releasing nutrients.

Besides, some dead organic matter lies in the mud, gradually reducing the margin of the pond, which is occupied by species whose leaves reach the water surface and roots remain in the mud. Gradually the water depth in the pond decreases due to evaporation and the deposition of organic matter, and the concentration of nutrients increases. Free floating plants increase in number because of the high nutrient availability. Gradually their dead parts fill up the pond bottom which gets raised. The pond becomes swampy ecosystem. The reed swamp species invade the pond and are gradually replaced by mesic communities as the water depth is reduced greatly. Gradually land plants invade. In association with the changes in water depth and vegetation, the aquatic fauna also changes and ultimately gets replaced by land animals.

Natural water resources may be found in some plateaus and valley. Hydrosere arising in these environments usually leads to the establishment of deciduous forests. Climax formation of trees like salix has been reported in low lying lands in some parts of Kashmir (Ambasht 1988). A possible trend in succession in the aquatic environment is an open scrub land. The process is as under:

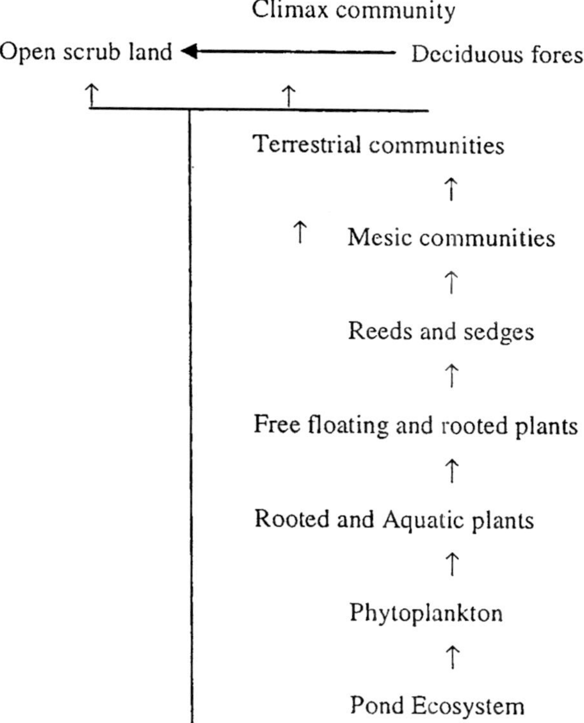

Xerosere or Xerarch Successio

This type of succession begins on exposed parent rocks (lithosere) or dry sand (psammosere). The original substratum is deficient in water and there is lack of organic matter, having only minerals in disintegrated un-weathered stage. The pioneer plants are lichens, mosses and selaginella, which help in soil formation by accelerating erosion. In course of time, grasses, annuals and herbacecus vegetation grows on soils deposited on rocks. Later the mixed woodland species appears which constitute the climax community. In xerosere also successive changes take place in both plants as well as animals. But as in most of the primary autotrophic successions, changes in plant life are more obvious than those in animals. The changes in plants are obvious to the extent that it looks as a succession of plant only and various stages are named on the basis of particular stage dominated by a particular plant species. Various stages in this succession are as under:

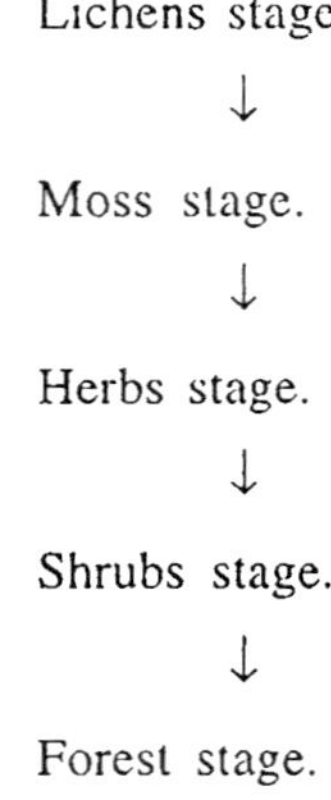

Thus in the xerosere the habitat changes from one of extreme to one of medium water relations and the vegetation, at first adapted to xeric conditions develop in to a mesophytic forest. The climax forest may be dominated by some sort of broad leaved plant community if the climate is conducive to the growth of such tree species or it may be characterized by alpine or sub alpine type tree species if the succession occurs at high elevations in the rocky mountains.

TRENDS IN SUCCESSION

While analyzing field examples and laboratory models, distinct structural and functional changes which occur in the process of succession are observed and have also been shown. Changes can be grouped as follows :

1. Species composition

A change in species composition occurs. The change occurs fast in the beginning and then more gradually.

2. Species diversity

Some plant species which were present in the initial stages may not be found in an advanced stage of succession. However, in the climax stage there may be more kinds of autotrophs than in the earlier serial stages.

3. Density and biomass of organisms

There is usually a marked increase in the number (density) of organism in older cultures or ecosystems. The number may decline in the older stages but in the climax stage the biomass structure remains very high. The total biomass gradually increases and reaches a maximum in the climax stage.

4. Heterotrophic population

The number of species usually goes on increasing, as the food chain relationships become more complex in the climax stage.

5. Chlorophyll

Green pigments go on increasing during the early phase of primary succession. The ratio of yellow/green pigments remains low in the early stages and increase in the climax stage. Pigment diversity also increases.

6. Functional Changes

(i) There is a progressive increase in the amount of living biomass and dead organic matter. There is an increase in grass as well as net primary production in the initial and serial stages. Thus there is more biomass accumulation, gradually reaching a huge biomass structure in the climax stage.

(ii) The community respiration increases but the P/R. ratio (community production/community respiration) remains more than 1 in the serial stage. The huge living biomass respires a lot in the climax and the P/R ratio equals 1 (P/R=1). Thus in the early stages P>R and in the climax stage P=R.

(iii) The food chain relationships becomes more complex as succession proceeds.

These field and laboratory examples indicate that the stability of the climax community is associated with high species diversity, large accumulation of living biomass and complex food chain relationships. The complexity of the climax community increases the number of ecological niches and of routes of energy flow through the system. These attributes make the climax community more stable.

CONCEPT OF THE CLIMAX

The final or stable community in a developmental series (sere) is the climax community. It is self perpetuating and in equilibrium with the physical habitat. Presumably, in a climax community, in contrast to a developmental or other unstable community, there is no net annual accumulation of organic matter. That is, the annual production and import is balanced by the annual community consumption and export. For a given region it is convenient, although somewhat arbitrary, to recognize (i) a single climatic climax, which is in equilibrium with the general climate, and (ii) a varying number of edaphic climaxes, which are modified by local conditions of the substrate. The former is the theoretical community towards which all successional development in a given region is trending, it is realized where physical conditions of the substrate are not so extreme as to modify the effects of the prevailing regional climate. Succession ends in an edaphic climax where topography, soil, water, fire, or other disturbances are such that the climatic climax cannot develop.

Many theories have been put forward to explain the climax concept :

The Mono Climax Theory

Clement (1916) was of the view that, in a given climate, the successional stages (serial stage) will ultimately end up as climatic climax vegetation. If this be true, succession is a progressive phenomenon. The emphasis is in the fact that only one type of climax vegetation develops. This is called the mono-climax theory. But some communities like prairies with grassland climax in the southern part of Canada and southern part of USA, possess shrubs and first patches in low area in the misic belt as stable vegetation. Ecologists supporting the mono-climax theory argue that this vegetation is post-climax. But others feel that the post climax concept is confusing. It has also been observed that different types of stable vegetation occur within the same climatic belt. Yet other plant ecologists consider this vegetation as pre-climax. Pre-climax vegetation is that which has not reached the climax stage due to the prevalence of some adverse climatic, topographic and edaphic environmental conditions (Misra 1974). Due to human activities, such as cutting, fire, grazing and so on, the successional process may be checked and the community may not reach the climax stage. However, some sort of stable vegetation may develop under such circumstances.

The Poly Climax Theory

Looking at the occurrence of several types of climax vegetation in Europe, Braun-Blanquet (1932) proposed the poly-climax theory, which states that there may be climatic climax, edaphic climax and biotic climax depending upon the situation in which the climax vegetation has developed. Therefore, the succession may not always be progressive. In course of succession, a community with large life forms like forest may be degraded and patches of grasslands may occur within them (retrogressive succession).

The two theories cited above explain the climax concept from the structural viewpoint. Ecologists have recently started looking at the problem from the energetic point of view, since energy derives all functions in a community or in the ecosystem. Researches in ecological energetics and related aspects have led to the development of another theory called the information theory, to explain the concept of climax community.

Information Theory

The community is considered a thermodynamic unit. It receives energy from the sun and converts it into chemical energy, performs its activities and dissipates heat energy. In the serial stages, the dissipation energy is usually less than the input energy-therefore, there is more of net production, and the community grows. In the stages, the species diversity is low and the food chain relationships become more complex and the possible interactions between individuals, species and materials increases. In the climax community the input energy more or less balances the output energy which makes for negligible net primary production. In short information theory takes energy parameters in to account and differentiates the serial stages or serial communities from the climax community. It becomes meaningful if we explain everything in terms of the energy budget, a functional approach to the study of ecological succession.

Time Factor in Ecological Succession

The establishment of a climax community through primary succession on sand dunes or recent lava flows take about 1000 years. Secondary succession on abandoned agricultural land or a cut-over forest site in a moist temperate climate and a tropical climate may take 200 and 100 years respectively for the establishment of a mature forest. Secondary succession in grasslands may take about 50 to 60 years to reach a climax grass stages. Odum (1966) describes four stages for abandoned croplands in the central and western North America to reach a climax grassland stages. These are (a) 2-6 years of annual weed stage, (b) 8-10 years of short lived grass sage, (c) 10-20 years of perennial grass stage and (d) the climax grasslands stage reached in some 20-40 years. In recent times, man and some natural events such as storm and climatic cycles, have interfered with the natural process of ecological succession. If a longer duration is required for completion of the sere, there is every likelihood of any of these factors interfering this processes. Therefore, long terms succession may not be directional and predictable.

SIGNIFICANCE OF ECOLOGICAL SUCCESSION

The principle and trends of ecological succession indicate that seral stages are more productive, although comparatively less stable. The climax community is mature and stable with greater biological diversity, larger biomass structure and balanced energy flow and is able to buffer the physical environment. This community provides man with food, fuel, fodder medicines and keep a balance with regard to biogeochmeical cycles. It is considered a multiple use system. Since the P/R rate in such a community is 1, there is not much net primary production for harvesting. The fast growing human population needs a huge amount of food and other materials. Therefore, man has to look for high net primary productive systems, which mean he must have early successional stages as a sources of food. Crop fields are highly productive systems because man puts in a lot of auxiliary energy into them, these system are comparable to successional stages. But, for human survival, a balance must be maintained between the stable, mature, climax system (life supporting systems) and the highly productive successional systems.

CHAPTER-X

BIODIVERSITY AND ITS CONSERVATION

The current popular term used for richness and diversity of life on this planet earth is Biodiversity or Biological diversity. Biodiversity means the wealth of life forms found on earth, in the form of millions of different plants, animals and microorganisms, which are further diversified with vast potential of future creation of the biodiversity- from the ocean of genomic diversity, at the effectively functional level within the living biomass, the genes they contain and the intricate systems they form.

DEFINITION

At the world convention for biological diversity, the following definition for biological diversity was recommended (UNEP 1992): ***"Biological diversity means the variability among living organisms from all sources including, interalia, terrestrial, marine and other aquatic ecosystems and the ecological complexes of which they are part; this includes diversity within species, between species and of ecosystems".*** Though this definition was fully accepted and adopted until 1995 all over the world, the definition in its make, context, concept and conformity altered and deviated variously and variedly (Balaji and Rai, 1999; Bhattacharya, 2000 and Khoshoo, 2000). These alienations infused confusion and fallacies both in its actual and basic concepts. Kour (2000) reported that 14 different definitions exist of diversity. Presently, biodiversity is defined as ***"the intrinsically-inbuilt plus the externally-imposed variability in and among living organisms existing in terrestrial, marine and other ecosystem at a specific period of time".***

GENETIC, SPECIES AND ECOSYSTEM DIVERSITY

The diversity includes the variability in the genes, genotypes, species, genera and families and ecosystems in a specific region at a specific time period. Inclusion of all the above given variables and addition of the interactions existent amongst these, renders the assessment of the overall biodiversity intricate. Therefore, the diversity of the above mentioned different components represented each at genetic, species, population and ecosystem levels needs to be assessed and evaluated individually at each single level at a specific period of time and thereafter may be considered on an overall and general basis. For instance, biodiversity at the genetic level represents the diversity of genes, which make up the germplasm of the species population. To this diversity, adds the diversity existing amongst various species that constitute the community. This species-diversity enhances the borders of biodiversity as it includes the variability existing within and between species and the interactions rampant amongst them. This harmonious existence and stability potential plus the genotype-environment interactions encapsulate them in a balance and sensitive system - the ecosystem.

The diversity existing among the various ecosystems in a region constitutes the ecosystem diversity. Hence, the interlocked interactions, both negative and positive existing amongst these three major levels of biodiversity, viz., the genetic, the species and the ecosystem, represent the general and total biodiversity of an area at a specific period of time. With the passage of time, the biodiversity changes due to the species-replacements, populations interactions and ecological succession. These changes are accentuated by the human-influence. Thus, the term biodiversity is therefore, used in the literature to cover both the number of different populations and species that exist and the complex interactions that occur among them. Biodiversity, therefore, is commonly considered at three different levels:

1. Within species (interaspecific) diversity; usually measured in terms of genetic differences between individuals or populations.

2. Species (interspecific) diversity; measured as a combination of number and evenness of abundance of species.

3. Community or ecosystem diversity; measured as the number of different species assemblages.

GENETIC DIVERSITY

Genes are the biochemical packages passed on by parents that determine the physical and biochemical characteristics of their offspring. Although, most of the genes are the same, subtle variations occur in some genes. The result of these variations may be obvious, such as size and color, or may be invisible, for example, susceptibility to diseases. Such genetic variability has made it possible to produce new breed of crops, plants and domestic animals, and in the wild allowed species to adapt to changing conditions.

SPECIES DIVERSITY

A group of organisms genetically so similar that they can interbreed and produce fertile offspring is called a species. Horses and Zebras are different species. They are genetically similar and can interbreed but all off springs are infertile. Species are usually recognizably different in appearance, allowing an observer to distinguish one from another, but sometimes the differences are extremely subtle. The species diversity is usually measured in terms of the total number of species within discrete geographical boundaries.

ECOSYSTEM DIVERSITY

The functional relationships within and among the communities and their environment are frequently complex, but they are the mechanisms of major ecological processes such as water cycle, soil formation, nutrient cycling and energy flow. These processes provide the sustenance required by living communities and so a critical interdependence results. Two different phenomena are frequently referred under the heading of ecosystem diversity: (a) the variety of species within different ecosystems- more diverse ecosystems contains more species; (b) the variety of ecosystems found within a certain biogeographical or political boundry.

Biodiversity is, therefore, an expression of both numbers and difference and can be seen as a measure of complexity (Gaston and Spicer 1998).

Biological diversity must be treated more seriously as a globel resource, to be indexed, used and above all, preserved. Three circumstances conspire to give this matter an unprecedented urgency include: (a) exploding human populations which is degrading the environment at an accelerating rate, especially in tropical countries, (b) science is discovering new uses for biological diversity in ways that can relieve both human sufferings and environmental destruction and, (c) much of the diversity is being irreversibly lost through extinction caused by the destruction of natural habitats.

Though nature and natural forces stabilize the biodiversity of an area, humans de- stabilize this by selective-exploitation, elimination, destruction of the species needed for use or not needed by them. Therefore, the biodiversity in an area, in a region, in a zone, in a country and in a continent is ever changing and is neither static, nor is nature controlled but is fully human-tailored. Hence, evaluation and assessment of biodiversity at species and at ecosystem levels becomes invalidated due to human invasion and influence. The biodiversity at the species level is also altered by selective retention and or cultivation of the species needed by humans and by the elimination of the species not of immediate use to humans. This alters the ecosystem biodiversity of an area as the species constitute the basic component of an ecosystem. Therefore, the biodiversity at the species-and at the ecosystem levels for an area is valid only at the time it is studied, evaluated and assessed, because humans constantly cause alteration in it.

BIOGEOGRAPHICAL REGIONS OF INDIA

Geological movements some 15 million years ago, joined the landmasses of India and Africa with Asia and Europe, resulting in the formation of Himalayan massif in Asia and the Alps and the Caucasus mountains in Europe. India is thus situated at the confluence of Afro-tropical, Euro-Asian and Indo-Malayan biogographic realms. ***Biogeography comprises the study of distribution, evolution, dispersal and environmental relationship of plants and animals in time and space.***

World has been divided in to following 6 zoo-giographical regions:

1. Neoarctic	North America and Green Land
2. Palaearctic	Eurasia excluding India, includes Iceland, Canary Islands, Korea, Japan and North Africa
3. Ethiopian	Africa south of Sahara
4. Oriental	India and Indo China
5. Australian	Australia, Ne Guinea and nearby islands
6. Neotropical	South and Central America, including the Antilles

According to the above zoogeographic classification, India forms a part of oriental region. Wallace (1876) classified Oriental region in to four sub regions:

(i) Indian Sub Region includes the geographical area in India.

(ii) Ceylones subregion

(iii) Indo-Chinese subregion

(iv) Indo-Malayan subregion

The Wild life Institute of India, Dehradun, in its report "Planning wildlife protected area net work in India" by Rodgers and Panwar (1988) has given a biogeographic classification of India. The classification recognizes ten biogeographic zones within which there are 25 biotic provinces. The biotic provinces have been further divided into sub-divisions or regions and the biomes: The biogeographic zones are as under:

1A: Trans Himalaya: Ladakh Mtns

1B: Trans Himalaya: Tibetan Plateau

2A: Himalaya: North West Himalaya

2B: Himalaya: West Himalaya

2C: Himalaya: Central Himalaya

2D: Himalaya: East Himalaya

3A: Desert: Thar

3B: Desert: Katchchh

4A: Semi Arid: Punjab plains

4B: Semi Arid: Gujarat, Rajasthan

5A: Western Ghats: Malabar plains

5B: Western Ghats: Western Ghats Mtns.

6A: Deccan Peninsula: Central Highlands

6B: Deccan Peninsula: Chotta Nagpur

6C: Deccan Peninsula: Eastern highlands

6D: Deccan Peninsula: Central Plateau

6E: Deccan Peninsula: Deccan South

7A: Gangetic Plains: Upper Gangetic Plains

7B: Gangetic Plains: Lower Gangetic Plain

8A: Coast: West Coast

8B: Coast: East Coast

8C: Coast: Lakshadweep

9A: North East: Brahamaputra Valley

9B: North East: North East Hills

10A: Islands: Andamans

10B: Islands: Nicobars

VALUE OF BIODIVERSITY

Humans have always relied on the natural world and, despite our technological development; this remains just as true today. Biological diversity existing on this planet has a great value. There is a multiplicity of ways of assessing value of biological resources. Economists, in particular, have tried various approaches but difficulty lies when applying a common formula to the variety of resources used by man. The value of forests in terms of logs, for example, would be measured in a different way from the value of forest for recreation or watershed protection. In the terms of Norton (1988) species can have value as commodities, and as amenities and they can also have moral values.

A species has commodity value if it can be made into a product that can be brought and sold in the market place.

A species has amenity value if its existence improves our lives in some non material ways e.g., the joy we experience in sightseeing, bird watching and the walks we enjoy in the middle of a forest.

Species have values as a moral resource to humans, as a chance for humans to form, re-form and improve their own value systems (Norton, 1984).

Mc Neely et al., 1990 have used three main approaches to determine the value of biological resources:

1. **CONSUMPTIVE VALUE**: the value of natural products e.g. firewood, fodder and game meat- these are consumed directly and do not pass through a market. A large variety of plants from wild are consumed as food. There are about 79,000 species of edible plants. Similarly a large number of wild animals are a source of food to us. Fuel wood is a sources of energy to us since man stated to cook his food. The fossil fuels which we are using at present are again a product of fossilized biodiversity.

2. **PRODUCTIVE VALUE**: the value of products that are commercially harvested, such as game meat sold in market, timber, ivory and medicinal plants. The important uses include silk from an insect, wool from sheep, lac from lac insect, honey from bees and meat from a wide variety of animals etc. which are traded in the market.

3. **NON-CONSUMPTIVE VALUE**: indirect value of eco-system functions, such as watershed protection,

photosynthesis, regulation of climate and soil production.

Than there are two more values added to biological resources:

1. **OPTION VALUE**: intangible values of keeping options open for the future. We have biological resources available around us which may one day prove to be an effective option for many other important aspects which may emerge in future.

2. **EXISTENCE VALUE**: Value attached to the ethical feelings of existence.

The consensus is that whatever methodology is used, valuation is only a fundamental first step. It informs planners, resource managers and local people about how important biological diversity may be to national development objectives. It further demonstrates how important an area is for the biological resources it contains, it reveals common interests in various sectors, and it facilitates comparison of costs and benefits of different development proposals. The second step is to determine how these species and areas can be conserved.

BIODIVERSITY AT GLOBAL, NATIONAL AND LOCAL LEVELS

Estimates of total number of species present on earth today range over more than an order of magnitude, from a low around 3 million to a high of 30 million or possibly much more. At the same time we have a little idea of the rate at which species are currently becoming extinct. The systematic naming and recording of species begin relatively recently, with Linneaus' work in 1758 recognizing some 9000 species (May, 2002). Recent studies indicate that about 1.4 million living species of organisms exist in this world. Approximately 750,000 are insects, 41,000 are vertebrates and 250,000 are plants. The remainder consists of a complex array of invertebrates, fungi, algae and micro organisms. Many taxonomists agree that this picture is still very incomplete except in a few well-studied groups such as the vertebrates and flowering plants. If insects, the most species rich of all major groups, are included, the absolute number is likely to exceed 5 million. Thus remarkably we do not know the exact number of species on the earth. According to May (2002) the total number of species that have been named and recorded emphasizes the uncertainties caused by the synonyms. The IUCN survey estimates that around 13,000 new species are currently named each year, but current rates of resolving synonymies reduces this number to around 10,000 distinct new species added yearly to be known total.

Table 10.1: Species Known To Exist (Adapted From Mcneel Et Al., 1990)

Group	No. Of described Species
Bacteria	4,760
Algae	26,900
Fungi	46,983
Mosses and Liverworts	17,000
Gymnosperms	750
Angiosperms	2,50,000
Protozoans	30,000
Sponges	5,000
Corals and Jellyfishes	9,000
Round worms and earthworms	24,000
Crustaceans	24,000
Insects	7,51,000
Other arthropods and minor invertebrates	1,32,000
Mollusks	50,000
Starfishes	6,100
Fishes	19,056
Amphibians	4,184
Reptiles	6,300
Birds	9,198
Mammals	4,170
Total	14,35,662

There is a wide variation on the amount of taxonomic efforts being carried out in identifying the unknown species in different groups. According to May (2002) roughly one third of all taxonomists working on vertebrates, another third are working on the ten times more numerous plant species, and the remaining third on invertebrate animals., which outnumber vertebrate species by at least a factor of 100.

According to May (1999), the current global total of distinct eukaryotic species that have been named and recorded is around 1.5 million. This is lower than Hammond's (1995) 1.74 million, but is consistent with Wilson's (1988) estimate of 1.4 million roughly 10 years ago. All such estimates are mainly dominated by total number of insects. The other factor is our ignorance of the ecological and evolutionary forces, which underpin these interesting but empirical species, size patterns.

Table 10.2: Number of named, distinct species of eukaryotes.

Group	Hammond (1995)	May (1999)
Protozoa	40,000	40,000
Algae	40,000	40,000
Plants	2,70,000	2,70,000
Fungi	70,000	70,000
Animals	13,20,000	10,80,000
(i) Vertebrates	45,000	45,000
(ii) Nematodes	25,000	15,000
(iii) Molluscs	70,000	70,000
(iv) Arthropods	10,85,000	8,55,000
(v) Arachnids	75,000	75,000
(vi) Insects	9,50,000	7,20,000
(vii) Others	20,000	20,000
Others	95,000	95,000
Total	17,40,000	15,00,000

At national level, India is known for its rich biodiversity which is again a characteristic feature of the country. The number of plant species in India is estimated to be over 45,000, representing about 7% of the world's flora which also includes over 15,000 species of flowering plants. The faunal wealth of India is equally rich and the total number of animal species is estimated to be 81,000, representing about 6.4 percent of the world's fauna.

Biodiversity at local and regional level is better understood by categorizing it in to following types based upon their spatial distribution:

(i) Alpha diversity: The diversity measured within a unit area e.g., a patch of wood land, a pond or a lake etc.

(ii) Beta diversity: This refers to the rate of change in species composition across different habitats e.g., change in species composition from Savana to forest or along an environmental gradient like altitudinal gradients.

(iii) Gamma diversity: This refers to the rate of change that occurs with increasing area.

INDIA AS A MEGA BIO-DIVERSITY NATION

Although most of the country is situated north of the tropic of cancer, India has a tropical monsoonal climate. It depends for rains on monsoon, which originates south of the Equator and on winter rain caused by western disturbances originating in the Mediterranean. India's immense biological diversity encompasses ecosystems, populations, species and their genetic make up. This diversity can be attributed to he vast variety in physiography and climatic situations resulting in a diversity of ecological habitats ranging from tropical, subtropical, temperate, alpine to desert. The Wildlife Institute of India has proposed a classification which divides the country into 10 bio geographic regions: Trans Himalayan, Himalayan, Indian Desert, Semi Arid, Western Ghats, Deccan Peninsula, Gangetic Plains, North East India, Islands and Coasts. India is one of the 12 identified mega biodiversity centers. It also has two of the 18 identified hot spots- the Eastern Himalayas and the Western Ghats.

The number of plant species is estimated to be over 45,000, representing about 7 percent of the world's flora. These are categorized in different taxonomic divisions including over 15,00 flowering plants. Estimates for the lower plants are: 64 gymnosperms, 2843 bryophytes, 1012 pteridophytes, 1940 lichens, 12,480 algae and 23,000 fungi.

Some 4,900 species of flowering plants are endemic to the country. Among the endemic species, 2,532 species are found in the Himalaya and adjoining areas, followed by 1,782 species in Peninsular India. About 1500 endemic species are facing varying degree of threat.

The faunal wealth is equally diverse. The total number of animal species is estimated at 81,000, representing about 6.4 percent of world's fauna. India's animal diversity includes over 5,000 molluscs and about 57,0000 insects, besides other invertebrates. There are 2,546 fish, 204 amphibians, 428 reptiles, 1228 birds and 372 mammals. The extent of endemism is very high, about 62 percent.

Very little is known about the extent of the diversity of micro-organisms, particularly the bacteria and viruses. Consolidated data on bacteria and viruses are to be compiled. Detailed survey has to be made about the aquatic biota related to the 14 major, 44 medium and 55 minor river systems.

India is also considered one of the world's 12 centers of origin of cultivated plants. India's rich germplasm resources includes 51 species of cereals and millets, 104 species of fruits, 27 species of spices and condiments, 55 species of vegetables and pulses, 24 species of fiber crops, 12 species of oil seeds, and various wild strains of tea, coffee, tobacco and sugarcane. There are several hundred species of wild crop relatives distributed all over the country especially in the Western and Eastern Himalayas, in the west Peninsula and the Malabar Coast, in North Eastern India, the Gangetic Plains and the Indus Plains. The ancient practice of domesticating animals has resulted in India's diverse livestock., poultry and other animal breeds. This is a significant percentage of world diversity. India's 8 breeds of buffalo represent the entire range of genetic diversity of buffalos in the world. Besides, there are 26 breeds of cattle, 40 of sheep, 20 of goats, 8 of camels, 6 of horses, 2 of donkeys and 18 types of poultry.

HOT SPOTS OF BIODIVERSITY

Biological diversity occurs within all habitats, because genetic diversity has allowed life to adapt to the harshest of environments. However, species are not spread evenly over the earth and biological diversity is greater in some areas than in others. Some habitats, particularly tropical forests among terrestrial systems, have a great number of species. Tropical forests cover only 7% of earth's land surface, yet they are estimated to contain at least 50% of all species.

Mittermeier (1988) and Mittermeier and Werner (1990) recognised that a very small number of countries situated mainly in the tropics possess a large fraction of the world's species diversity, and introduced the term of " Megadiversity Countries", which they suggested, merit special international attention. McNeel et al., (1990) used country species lists of vertebrates, swallowtail butterflies and higher plants to identify 12 such megabiodiversity countries (Fig.1): ***Mexico, Colombia, Equador, Peru, Brazil, Zaire, Madagaskar, China, India, Malaysia, Indonesia,*** and ***Australia.*** Together these countries hold upto 70% of the world's species diversity in these groups. This approach is relatively simple in that it involves species inventory within a given geopolitical boundary. It also recognises that conservation action is managed at country level. There may be considerable overlap in species composition between different regions with high species numbers, particularly if they are situated close to one another geographically. For example, in terms of mammal species in megadiversity countries Equador with 271 species and Peru having 344 species, 208 species are common to both the countries. In addition high diversity countries contain large numbers of very widely distributed species, which are currently neither threatened nor otherwise of species conservation concern.

There are a total 18 hot spots identified throughout the globe which are as follows:

1. Colombian Choco
2. Western Ecusdor
3. Uplands of western Amazonia
4. Atlantic forest region of eastern Brazil
5. Eastern Madagascar
6. Peninsular Malaysia

7. Northern Borneo
8. Queens land Australia
9. New Caledonia
10. California floral province
11. Central Chile
12. Ivory Coast
13. Cape floral province
14. Southwest Australia
15. Eastern Arc forests Tanzania
16. Western Ghats
17. Sri Lanka
18. Southwest Australia

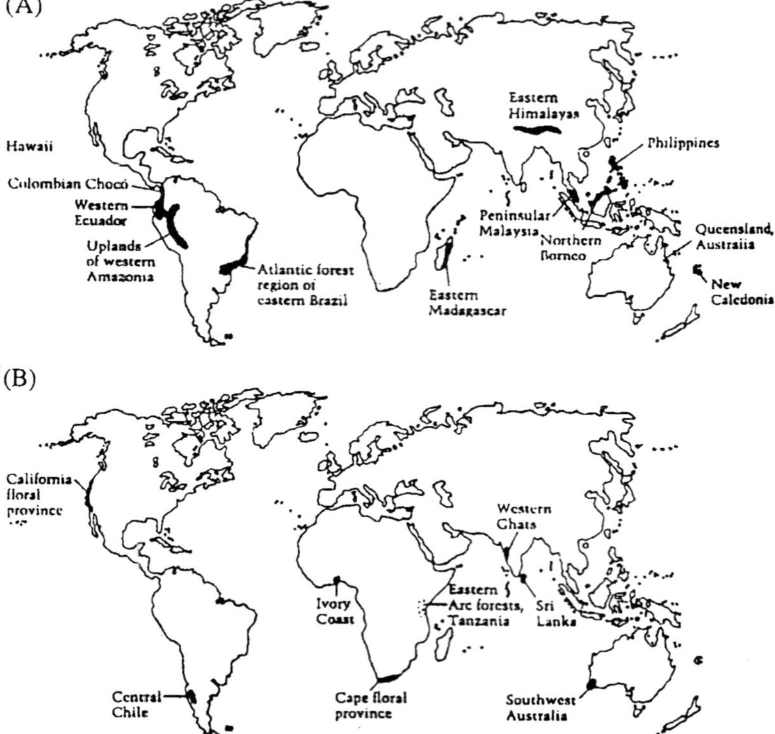

Fig 1. A- 12 Tropical rainforest hotspots, B- Hotspots in other climatic regions

THREATS TO BIODIVERSITY

The global loss of biological diversity has been described as a product of two phenomenon:

1. Population levels have forced the transformation of relatively undisturbed areas into lands used for agriculture, and

2. Both industrial and agricultural pollutants have applied a new and narrowly uniform selective pressure on species.

Population growth and technological changes have multiple, rather than simply additive impacts on biological diversity. However, from a conservationists point , the various agents responsible for decline in biodiversity are classified under four headings:

1. Habitat loss

2. Poaching of wild life

3. Impact of introduced species

4. Chains of extinction.

Dimond (1984) called these "the evil quarter". From an analysis of factors responsible for known and threatened extinctions of vertebrates, Jenkins (1992) has shown that introduced animals and direct habitat destruction are major contributors. 39% of species have become extinct through introductions and 36% through habitat loss. Hunting and deliberate exterminations contribute significantly (23%) of extinctions with known causes.

1. Habitat Loss

The most direct threat to biodiversity comes from destruction of the habitat on which it depends. Although habitats may be modified, degraded or merely eliminated, it is more commonly fragmented. A large tract is converted piecemeal into another land use. This process is common throughout the world, as can be seen not only in India but also in many countries specially the developing countries. Loss of habitat by a given proportion does not increase the vulnerability of a species, not decreases the numbers of its members, but that same proportion, except in the special case of habitat being cleared from the edge inward. More commonly the modification produces a patchwork of pattern as it erodes the tract of habitat from inside and changes microclimate. At first the areas occupied by the new land use provides the continuous phase and the original habitat the discontinuous one. The vulnerability of species then increases disproportionately.

In developing countries, the patterns and causes of habitat destruction are linked to economic growth. In economically weak developing countries rapid population growth causes economic and political problems that require resources. These are provided at the expense of the environment, mainly through forest clearance. Ehrlich (1995) develops the argument that habitat destruction is linked to human enterprise, which in turn is directly correlated with energy usage. Therefore, if habitat destruction is a major cause of species extinction, both of these will increase as the human population grows.

2. Poaching of Wild Life

Overkill results from hunting at a rate above maximum sustained yield. The most susceptible species are those with low intrinsic rate of increase, because of their limited ability to recover quickly. Hence, although usually having a high standing biomass when unharvested, they give a low maximum sustained yield, which is easily exceeded. They are even more vulnerable if valued either as food or as an easily marketable commodity. The harvesting of wild plants and animals is still a major human activity. As humans became more numerous and wide spread they started to kill species faster than those species could replace themselves, either through reproduction or immigration from else where. Population will crash if we continue to remove natural resources at a greater rate than they can be replaced. Thus the vulnerability of species to overexploitation is partly predicted by their capacity to increase. Apart from the direct threat of overkill of a species, there are indirect consequences for the communities and ecosystems to which these species belong. All species play definite role within ecosystems. However, some species play a unique and important role in ecosystem function and their removal results in fundamental change in that system.

3. Impact of Introduced Species

Species intentionally or unintentionally introduced by people have produced to exterminate native species by competing with them, preying upon them, or destroying their habitat. A review by Atkinson (1989) has demonstrated that 22 species and subspecies of reptiles and amphibians have disappeared worldwide due to alien animals. In New Zealand alone, 9 species of reptiles and amphibians and 23 bird species have become extinct since AD 1000 through introduction. The introduction of exotic species into some areas has had a devastating impact on the native biodiversity. Native species most vulnerable to the impact of exotic species introduction are those, which have evolved in isolation from high levels of competition and predation. The majority of recent extinct species inhabited small isolated oceanic island. Some of these, such as the Dodo, were early targets for sailors for much needed protein, but many more smaller flightless birds, were victims of species introduced by man, accidentally or deliberately. There are a number of characteristics of species that make them good invaders:

(i) Species with a high reproductive rate that can quickly build up a large population size under favorable conditions.

(ii) Generalists species with broad habitat requirements and diet that are likely to be able to find somewhere suitable to live and something suitable to eat.

(iii) Good dispersers that can travel across the new area and find suitable habitats.

Similarly there are a number of characteristics of places that may make it easier for introduced species to become established:

- Early successional areas or disturbed areas with empty niches tend to have resources that are unexploited and available for immigrating species with little or no competition.
- Remote islands with low taxonomic diversity are vulnerable because few species have been able to colonize and communities and food webs are often simple with potential empty niche space to be exploited.
- Remote islands with no predators and native prey providing an easily exploited food source.

4. Chains of Extinctions

The extinction of one species may cause the demise of others, this is known as secondary extinctions. Diamond (1989) gave as example the extinction or near extinction of genus Hibiscadelaphus as a consequence of the extinction of several species of pollinators, the Hawaiian honeycreepers. Examples of predators and scavengers dying out after the species providing their food died out are also common.

MAN WILDLIFE CONFLICT

The loss of natural habitats has led in the recent times a lot of conflict between man and wild life. We have been reading reports in the newspapers regarding the destruction of crops, huts and other materials by the different species of animals. During 2004-05 as many as 12 children died in the hills of Uttarakhand due to the attack of leopard. The instances were more in Uttarkashi and Pauri districts. Similarly damage to agricultural crops by elephants and neelgai in the district of Haridwar is a major problem. Occasionally Elephants have been causing loss to human life in this part of the state. Such reports are also available from other parts of country e.g., In Orissa about 200 people were killed by elephants during 2001-05, villagers have also killed many elephants and injured as many as 30 elephants.The death of elephants in the railway tracks in the Rajaji National park was an International News in the year 2003. Such instances keep on happening nowand then.

Man animal conflicts have also been reported from Kaziranga National Park, Corbett, Dudhwa, Palamau and Ranthambore National Parks in the Country. In 2004 and also before that instances of entering of leopards in the IIT Campus in Powai in Mumbai were came into light which created a lot of panic among the residents in the area.

IUCN CATEGORIES OF SPECIES

The following categories of rare plants and animals have been recognized by the International Union for Conservation of Natural Resources (IUCN), mainly based on (I) present and past distribution, (ii) decline in number of populations in course of time, (iii) abundance and quality of natural habitats, and (iv) biology and potential values of the species. These are defined as follows for the purpose of conservation:

1. Endangered (E)

This includes the species which is in danger of extinction and whose survival is unlikely if the casual factors continue operating. Included are the taxa whose numbers have been reduced to a critical level or whose habitats have been so drastically reduced that they are deemed to be in immediate danger of extinction.

2. Vulnerable

This includes the taxa believed likely to move into the endangered category in the near future if the casual factors continue operating. Included are the taxa, of which most or all the populations are decreasing because of overexploitation, extensive destruction of habitat or other environmental disturbances; taxa with populations that have been seriously depleted and whose ultimate security is not yet assured; and taxa with populations that are still abundant but are under threat from serious adverse factors throughout their range.

3. Rare (R)

It includes the taxa with small world populations that are not at present endangered or vulnerable, but are at risk. These taxas are usually localized within restricted geographical areas or habitats or are thinly scattered over a more extensive range.

4. Threatened (T)

The term threatened is used in the conservation context for species, which are in one of the categories of endangered, vulnerable and rare. Some species are marked as threatened where it is known that they are endangered, vulnerable or rare, but there is not enough information to say which of these three categories is appropriate.

5. Out of Danger (O)

It includes the taxa formerly included in one of the above categories, but which are now considered relatively secure because the effective conservation measures have been taken or the previous threats to their survival have been removed.

6. Intermediate (I)

This includes the taxa that are suspected of belonging to one of the first three categories, but for which insufficient information is currently available.

ENDANGERED AND ENDEMIC SPECIES OF INDIA

According to a recent report of IUCN in the year 2000, a total of 215 animal species are threatened including 86 mammals, 70 birds, 25 reptiles, 3 amphibians, 8 fishes, 2 molluscs and 21 other invertebrates. Schedule 1 of the Wildlife (Protection) Act, 1972, gives a list of threatened species of India, which is given as below:

MAMMALS

1. Andaman wild pig (sus andamanensis)
2. Bharal (Ovis nahura)
3. Binturong (Arctictis binturong)
4. Brow-antlered deer or thamin (Cervus eldi)
5. Himalayan brown bear (Ursus arctos)
6. Capped langur (Presbytis pileatus)
7. Caracal (Fielis caracal)
8. Cheetah (Acinonyx jubatus)
9. Chinese pangolin (Manis pentadactyla)
10. Chinkara or Indian gazelle (Gazella gazella bennetti)
11. Clouded leopard (Neofelis nebulosa)
12. Crab eating macaque (Macaca irus umbrosa)
13. Desert cat (Felis libyca)
14. Desert fox (Vulpes bucopus)
15. Dugong (Duong duong)
16. Ermine (Mustela erminea)
17. Fishing cat (Felis viverrina)
18. Four horned antelope (Tetraceros quadricornis)
19. Gangetic dolphin (Platanista gangetica)
20. Gaur or Indian biason (Bos gaurus)

21. Golden cat (Felis temmincki)
22. Golden langur (Presbytis geei)
23. Giant squirrel (Ratufa macroura)
24. Himalayan ibex (Capra ibex)
25. Himalayan tahr (Hemitragus jemlahicus)
26. Hispid hare (Caprolagus hispidus)
27. Hog badger (Arctonyx collaris)
28. Hoolock gibbon (Hylobates hoolock)
29. Indian elephant (Elephas maximus)
30. Indian lion (Panthera leo persica)
31. Indian wild ass (Equus hemionus)
32. Indian wolf (Canis lupus pallipes)
33. Kashmir stag (Cervus elaphus hanglu)
34. Leaf monkey (Presbytis phayrei)
35. Leopard or Panther (Panthera pardus)
36. Leopard cat (Felis bengalensis)
37. Lesser or Red panda (Ailurus fulgens)
38. Lion tailed macaque (Macaca silenus)
39. Loris (Loris tardigradus)
40. Little Indian porpoise (Neomeris phocaenoides)
41. Lynx (Felis lynx isabellinus)
42. Malabar civet (Viverra megaspila)
43. Sun Bear (Helarctos malayanus)
44. Marbled cat (Felis marmorata)
45. Markhor (Capra falconeri)
46. Mouse deer (Tragulus meminna)

47. Musk deer (Moschus moschiferus)
48. Nilgiri langur (Presbytis johni)
49. Nilgiri tahr (Hemitragus hylocrius)
50. Great Tibetan sheep (Ovis ammon hodgsoni)
51. Pallas's cat (Felis manul)
52. Pangolin (Manis crassicaudata)
53. Pygmy hog (Sus salvanius)
54. Ratel (Mellivora capensis)
55. Indian onc horne rhinoceros (Rhinoceros unicornis)
56. Rusty spotted cat (Felis rubiginosa)
57. Serow (Capricornis sumatraensis)
58. Clawless otter (Aonyx cinereal)
59. Sloth bear (Melursus ursinus)
60. Snow leopard (Panthera uncia)
61. Small Travancore flying squirrel (Petinomys fuscopapillus)
62. Sunfin dolfin (Orcaella brevirostris)
63. Spotted linsang (Prionodon pardicolor)
64. Swamp deer (Cervus duvauceli)
65. Takin (Budircas taxicolor)
66. Chiru (Panthilops hodgsoni)
67. Tibetan fox (Vulpes ferrilatus)
68. Tibetan gazelle (Procapra picticaudata)
69. Tibetan wild ass (Equus hemionus kiang)
70. Tiger (Panthera tigris)
71. Urial or Shapu (Ovis vignei)

72. Wild buffalo (Bubalus bubalis)
73. Wild yak (Bos grunniens)
74. Tibetan wolf (Canis lupus chanco)

Part-II

Amphibians and Reptiles

1. Audithia turtle (Pelochelys bibroni)
2. Yellow monitor lizard (Varanus flavescens)
3. Crocodiles (Crocodilus porosus and Crocodilus palustris)
4. Terrapin (Batagur baska)
5. Eastern hill terrapin (Melanochelys tricarinata)
6. Gharial (Gavialis gangeticus)
7. Ganges shoft shelled turtle (Trionyx gangeticus)
8. Golden gecko (Calodactyloides aureus)
9. Green sea turtle (Chelonia mydas)
10. Hawk sbill turtle (Eretmochelys imbricata imbricata)
11. Indian egg eating snakes (Elachistodon westermanni)
12. Indian shoft shelled turtle (Lissemys punctata)
13. Indian tent turtle (Kachuga tecta tecta)
14. Kerala forest terrapin (Hoesemys sylratiea)
15. Leathery turtle (Dermochelys coriacea)
16. Loggerhead turtle (Caretta caretta)
17. Oliveback loggerhead turtle (Lepidochelys olivacea)
18. Peacock marked soft shelled turtle (Trionyx hurum)
19. Pythons (Genus pythons)
20. Sail terrapin (Kachuga kachuga)
21. Spotted black terrapin (Geoclemys hamiltoni)

Part- III

Birds

1. Andaman teal (Anas gibberifrons albogularis)
2. Assam bamboo partridge (Bambusicola fytchii)
3. Bazas (Aviceda jerdoni and Aviceda leuphotes)
4. Bengal florican (Eupodotis bengalensis)
5. Black necked crane (Grus nigricollis)
6. Blood pheasants (Ithaginis cruentus tibetanus)
7. Cheer pheasant (Catreus wallichi)
8. Eastern white stork (Ciconia cikonia boyciana)
9. Forest spotted owlet (Athene blewitti)
10. Frogmouths (Genus batrachostomus)
11. Great Indian bustard (Choriotis nigriceps)
12. Great Indian Hornbill (Buceros bicornis)
13. Hawks
14. Hooded crane (Grus monacha)
15. Hornbills
16. Houbara bustard (Chlamydotis undulata)
17. Hume's bar backed pheasant (Syrmaticus humiae)
18. Indian pied hornbill (Anthracoceros malabaricus)
19. Jeordon's courser (Cursorius bitorquatus)
20. Lammergeier (Gypaetus barbatus)
21. Large falcons (Falco peregrinus)
22. Large whistling teal (Dendrocygna bicolour)
23. Lesser florican (Sympheotides indica)
24. Monal phesant (Lophophorus impejanus, L. sclateri)

25. Mountain quail (Ophrysia supercıliosa)
26. Narcondam hornbill (Rhyticeros narcondami)
27. Nicobar megapode (Megapodius freycinet)
28. Nicobar pigeon (Caloenas nicobarica pelewensis)
29. Fish eating eagle (Pandion haliactus)
30. Peacock pheasants (Polyplectron bicalcaratum)
31. Peafowl (Pavo cristatus)
32. Pink headed duck (Rhodonessa caryophyllacea)
33. Scalater's monal (Lophophorus sclateri)
34. Siberian white crane (Grus leucogernaus)
35. Tibetan snow cock(Tetraogallus tibetanus)
36. Tragopan pheasants (Tragopan melanocephalus)
37. White bellied sea eagle (Haliactus leucogaster)
38. white eared pheasant (Crossoptilon crossoptilon)
39. White spoonbill (Plate lea leucorodia)
40. White winged wood duck (Cairina scutalata)

PART-IV

Cruatacea and Insects

Butterflies and Moths

Family Amathussidae

Discophora deo deo
Discophora sondaica muscina

Faunis faunula fauniloides

Family Danaidae Danaus gautama gautamoides

Euploea crameri nicevillei

Euploea midamus roepstorfti

Family Lycaenidae

Allotinus drumila
Allotinus fabius penormis
Amblopala avidiena

Amblypodia ace arata
Amblypodia alea constanceae
Amblypodia ammon ariel

Amblypcdia arvina ardea
Amblypodia asopia
Amblypodia comica

Amblypodia opalina
Amblypodia zeta
Biduanda melisa cyana

Callophyrs leechii
Castalius rosimon alarbus
Charana cepheis

Chlioria othona
Deudoryx epijarbs amatius
Everes moorei

Gerydus biggsii
Gerydus symethus diopeithes
Heliophorus hybrida

Jamides Ferrari
Liphyra brassolis
Listeria dudgeni

Logania wastiniana subfaciate
Lycaenopsis binghami

Lycaenopsis haraldus ananga
Lycaenopsis purpa prominens

Lycaenopsis quadriplaga dohertyi
Nacaduba noreia hampsonii
Polymmatus orbitulus leela
Pratapa iocetas mishmia

Simiskina phalena harterti
Sinthusa virgo

Spindasis rukmini
Strymonidia mackwoodi
Tajuria ister

Tajuria luculentus nela
Tajuria yajna yajna
Thecla bieti menlera

Thecla ataxus zulla
Thecla letha

Thecla paona

Thecla pavo Virachola smiles

Family Nymphalidae

Apatura ulupi ulupi Argynnis hegemone
Calinaga Buddha

Charaxes durnfordii nicholi
Cirrochroa fasciata

Diagora nicevillei Dilipa morgiana
Doleschallia bisaltide andamana

Eriboea schreiberi Eulaceura manipurensis
Euthalia durga splendens

Euthalia iva Euthalia khama curvifascia
Euthaliatelchnia

Helcyra hemina Hypolimnas missipus
Limenitis austenia purpurascens Limenitis zumela
Melitaea shandura

Neptis antilope Neptis aspasia
Neptis columella kankena

Neptis cydippe kirbariensis
Neptis ebusa

Neptisjumbah binghami
Neptis manasa

Neptis nyctens Neptis Poona
Neptis sankara

Reta moorei Prothoe franckii regalis Sasakia
funebris

Sephisa chandra Symbrenthia silana
Vanessa antiopa yednula

Family Papilionidae

Chilasa clytia clytia f. commixtus
Papilio elephenor

Papilio liomedon — Parnassius acco geminifer

Parnassius delphius — Parnassius hannyngtoni
Parnassius imperator augustus

Parnassius stoliczkanus — Polydorus coonsambilanga

Polydorus crassipes — Polydorus hector
Polydorus nevilli

Polydorus plutonius pembertoni
Polydorus polla

Family Pieridae

Aporia harrietae harrietae — Baltia butleri sikkima

Colias colias thrasibulus — Colias dubi
Delias sanaea

Pieris krueperi devta

Family Satyriidae

CoElites nothis adamsoni — Cyllogenes janetae
Elymnias peali

Elymnias penanga philansis — Erabia annada annada
Erabia narasingha narsingha
Lethe distans — Lethe dura gammiee

Lethe europa tamuna — Lethe gemina gafuri
Lethe guluihal guluihal

Lethe margaritae — Lethe ocellata lyncus
Lethe ramadeva

Lethe satyabati — Mycalesis orseis nawtilus

Parargemenava macroides — Yothima dohertyi persimilis

The number of plant species in India is estimated to be over 45,000 representing about 7 percent of the world's flora. These are categorized into different taxonomic divisions including over 1500 flowering plants, 2843 bryophytes, 1012 pteridophytes, 1940 lichens, 12,480 algae and 23,000 fungi. Some 4,900 species of flowering plants are endeimic to the country this includes 2,532 species found in the Himalayan region and 1,782 species in Peninsular India. About 1500 endemic species are facing varying degree of threats.

CONSERVATION OF BIODIVERSITY

There are two good reasons for conserving biodiversity. The first is moral: it is right to do so. The second is practical: biological diversity supports human survival, notably through health, food and industry. The fundamental social, ethical, cultural and economic values of biological diversity have been recognized in most of the human disciplines, from religion to science. Given these multiple values it is not surprising that most cultures consider the importance of conservation. Despite this and in order to compete for the attention of the government decision makers throughout the world, policies regarding the protection of biological diversity must embrace economic values.

The concern for Biological Diversity is, however, a concern for man himself. All forms of life- human, animal and plants, are so closely interlinked that disturbance in one gives rise to imbalance in the others. If species of plants and animals become endangered they signify degradation in the environment, which may threaten man's own existence.

EX-SITU CONSERVATION

Ex-situ conservation means "conservation outside the habitats by perpetuating sample population in genetic resources centers, zoos, botanical gardens, culture collections etc. or in the form of gene pools and gamete storage for fish; germ plasm banks for seeds, pollen, semen ova, cells etc. This form of conservation includes the following:

1. Zoos

Zoo or zoological garden is a place where wild animals are kept for public showing. Some of the zoos have rare animals. They have recorded success with captive breeding of animals. For example, the Pere David's Deer and the European Bison, having lost their natural habitats many decades ago, have been kept alive solely through zoos and similar facilities.

2. Botanical gardens

They play an important role in the conservation of plant species so much so that there are several instances when plants believed to be extinct were fond living only in a botanical garden. Sophora toroniro is the famous example. The Botanical Survey of India (BSI) with its main botanic garden at Sibpur and the regional experimental gardens established at different places to suit the acclimatization of plants occurring in different ecological condition in the country have in their holding rare, endemic and threatened plants (Jain and Sastry, 1980). Records of threatened plants that are in cultivation have been kept in "Green Books". The Indian Greek Book prepared by growing in living state in some botanic gardens of the BSI in the country. In the present day context, botanic gardens have assumed a special role as center where active research and action oriented plan for conservation can be developed.

3. Seed Banks

Such conservation is practiced through cold storage in seed banks where seed is stored for long durations. Even after germ plasm of primitive cultivars or land races and other cultivars is introduced in seed banks, the possibility of diversity being lost in still there on account of the fact the seed samples are often not large enough to cover the whole spectrum of variation. There is bound to be shrinkage of breeding groups and hence of the gene pools.

IN SITU CONSERVATION

At present we have 12 biosphere reserves, 87 National parks, 485 wildlife sanctuary and 120 botanical gardens in our country covering 4% of geographical area.

The Biosphere Reserves conserve some representative ecosystems as a whole for long term in-situ conservation. In India we have 12 biosphere reserves viz. Nadadevi (Uttrakhand), Nilgiri (Karnataka, Kerala & TN), Nokrek (Meghalaya), Dibru Saikhowa and Manas (Assam), Sunderbans (W. Bengal), Gulf of Mannar (TN), Great Nicobar (Andaman and Nicobar), Simplipal (Orissa), Dehang Debang (Arunachal Pradesh), Pachmari (MP) and Kaziranga (Sikkim).

Within Biosphere Reserve we may have one or more National Parks. For e.g. Nilgiri has two National Parks viz Bandipur and Nagarholi National Park.

National Park is an area dedicated for the conservation of wild life along with its environment. It is also meant for enjoyment through tourism but without disturbing the environment. All other activities are prohibited within a National Park. Some major national parks are Rajaji and Corbett National Parks (Uttarakhand), Kaziranga (Assam), Gir National Park (Gujrat), Dachjigam (J&K), Periyar (Kerala), Kanha (M.P.), Dudhwa (UP) and Ranthambore and Sariska (Rajashthan).

Wildlife sanctuaries are also protected areas where killing, hunting, shooting or capturing of wildlife is prohibited. However, private ownership rights are permissible and forestry operations are also permitted to an extent that they do not affect the wildlife adversely. Some major wildlife sanctuaries are Ghana Bird Sanctury (Rajashtan), Hazribagh Sanctury (Bihar), Wild Ass Sanctury (Gujrat), Mudamalai (TN), Jaldapara wildlife sanctuary (W. Bengal)

Similarly some projects were started by the Government of India to conserve the endangered species within their natural habit and habitat e.g., Project Tiger (1973); Project Elephant (1992); Crocodile breeding Project (1975); Project Hangul (1970); Conservation of Himalayan Musk Deer (1974); and The Gir Lion Sanctuary Project (1972).

CHAPTER-XI

Environmental Pollution

The term pollution may be defined as ***"Addition of any foreign material (inorganic, biological or radiological) or any physical change in the natural environment which may adversely affect the living resources directly or indirectly, immediately or after sometime or after a very long time".*** Pollution may also be defined as , "the unfavorable alteration of our surroundings, wholly or largely as a by product of man's action." One simple definition of pollution is "An undesirable change in the physical, chemical or biological characteristics of air, water and soil that may harmfully affect the life or create a potential health hazard for any living organism". Pollution is thus direct or indirect change in any component of biosphere that is harmful to the living component (s), and in particular undesirable for man, affecting adversely the industrial progress, cultural and natural assets or general environment.

Every organism in a natural ecosystem produces potentially polluting waste products. What makes natural ecosystem sustainable is that the waste from one kind of organism becomes food and or raw material for another. In balanced ecosystems, waste do not accumulate to produce unfavorable alterations, they are broken down and recycled.

Through much of their history humans have relied on the same natural process to dispose of their wastes. But the situation has become extremely unbalanced. Exploding human population coupled with increasing use of materials and energy have led to enormous volumes of wastes and other materials being discharged in to the environment. Even when materials are biodegradable, that is, of a kind that can be assimilated and recycled by organisms, sheer volumes overwhelm the capacity of natural system to cope.

Aggravating the problem is the production of increasing amounts and kinds of non-biodegradable materials, which are not readily broken down and assimilated by the natural processes.

India today is one of the first ten industrialized countries of the world. Today we have a good industrial infrastructure in core industries like metals, chemicals, fertilizer, petroleum, food etc. What has come out of these? Pesticides, detergents, plastics, solvents, fuels, paints, dyes, food additives etc. are some examples. Due to progress in atomic energy, there has also been an increase in radioactivity in the biosphere. Besides these, there are a number of industrial effluents and emissions particularly poisonous gases in the atmosphere. Mining activities have also added to this problem particularly as solid waste.

Clearly, pollution involves so many different factors from so many sources that there is no single or simple remedy. Pollution is an evil which has born out of development. Lack of development of a culture of pollution control, has resulted a heavy backlog of gaseous, liquid and solid pollution in our country. It is to be cleaned. Thus pollution control in our country is a recent environmental concern. Not only in India, but also in the undeveloped Western World, pollution is a scare-word, which is a man-made problem, mainly of affluent countries. The developed countries have been in a mad race to exploit every bit of natural resource to convert them into goods, for their comfort, and to export them to needy developing world. In doing so, the industrialized countries dump lot of materials in their environment which becomes polluted. In one way pollution has been in fact "exported" to developing countries.

POLLUTANTS

Substance which are responsible for a change in the natural environment or conditions, which are harmful to the nature or particularly to the human beings, are called pollutants. A pollutant may thus include any chemical or geo-chemical (dust, sediment, grit etc.) substance, biotic component or its product, or physical factor (heat) that is released intentionally or unintentionally by man into the environment in such a concentration that may have adverse harmful or unpleasant effects. A pollutant has also been defined as ***"any solid, liquid or gaseous substance present in such concentration which may be or may tend to be injurious to the***

environment". Pollutants are the residues of things we make, use and throw away. There are many sources of such pollutants. As the clean air moves across the earth surface, it collects the products of both natural events such as dust storms, volcanic eruptions etc. and human activities such as vehicular emissions etc. These potential pollutants, called primary pollutants, mix with the churning air in the troposphere. Some may react with one another or with the basic components of air to form new pollutants called secondary pollutants. Long lived pollutants can travel great distances before they return to the earth surface as solid particles, droplets, or chemicals dissolved in precipitation.

The following pollutants have been identified as most wide spread and serious:

(i) **Particulate**: Particulate are tiny solid or liquid particles suspended in the air. These particles can be seen as smoke or haze. Other pollutants present as gas or vapor are not visible except in the case of nitrogen oxide, which is a brownish gas. Particulate may carry any or all of the other pollutants dissolved in or adhering to their surfaces.

(ii) **Hydrocarbons and other volatile organic compounds**: These include materials such as gasoline, paint solvents, and organic cleaning solutions that evaporate and enter the air in a vapor state.

(iii) **Carbon monoxide**: This is a highly poisonous gas.

(iv) **Nitrogen oxides**: These include several nitrogen oxygen compounds, all in gaseous forms.

(v) **Sulfur oxides**: Sulfur di-oxide is a poisonous gas to both plants and animals.

(vi) Lead and other heavy metals.

(vii) Ozone and other photochemical oxidants.

(viii) Acid droplets.

(ix) **Metals:** Mercury, lead, iron, zinc, nickel, tin, cadmium, chromium etc.

(x) **Agrochemicals**: Pesticides, herbicides, fungicides, nematicides bactericides, weedicides and fertilizers etc.

(xi) Various effluents and solid wastes which are not treated prior to their discharge.

(xii) Radioactive waste.

On the basis of their being bio-degradable or non- degradable nature the pollutants may be grouped in one of the following two categories:

1. Non-degradable pollutants

These are the materials and poisonous substances like aluminum cans, mercuric salts, long-chain phenolics, DDT etc. that either do not degrade or degrade only very slowly in nature. They are not cycled in ecosystem naturally. They not only accumulate but also are often biologically magnified with their subsequent movement in food chains and biogeochemical cycles.

2. Biodegradable pollutants

They are the domestic wastes than can be rapidly decomposed under natural conditions. They may create problems when they accumulate (i.e. their input into the environment exceeds their decomposition).

In United Nations Environment Program (UNEP) document following order of priority of different pollutants has been indicated:

S.N.	Order of priority	Medium
1.	SO_2 + suspended particle, strontium, cesium.	Air and food
2.	Ozone DDT and other organochlorine compounds	Air Biota, Man
3.	NO_3s. No_2s Nitrogen oxides	Drinking water Air
4.	Mercury compounds Lead and cobalt	Food, Water Food, Air
5.	Petroleum hydrocarbons Carbon monoxide	Sea Air
6.	Fluorides	Water (fresh water)
7.	Asbestos Arsenic	Air Drinking Water
8.	Mycotoxins and microbial contaminants	Food

TYPES OF POLLUTION

Pollution may be classified on the basis of the type of environment that is being polluted. Thus we may have air pollution, water pollution and soil or land pollution. Another classification of pollution may be on the basis of type of pollutants such as sulfur dioxide pollution, fluoride pollution, carbon monoxide pollution, smoke pollution, lead pollution, mercury pollution, solid waste pollution, radioactive pollution, noise pollution etc. Organisms do have the capacity to deal with certain amount or level of pollutants without suffering ill effects. The level of a pollutant below which no ill effects are observed is called the threshold level. Above the threshold level effects begin to be observed. However, the effects caused by a pollutant depend on its concentration and time of exposure.

AIR POLLUTION

Normal composition of clean air at or near sea is as given below :

Gases	Percent (by volume)
Nitrogen	78.084
Oxygen	20.9476
Argon	0.934
Carbon dioxide	0.0314
Methane	0.0002
Hydrogen	0.00005
Other gases	minute

However, it is very difficult to find such clean air and the concentration levels of different gases show variations due to addition of different types of pollutants in to the air which may be in the form of gases or particulates making it unfit for living organisms.

CAUSES OF AIR POLLUTION

In large measures, air pollutants are direct and indirect byproducts of burning coal, gasoline and other liquid fuels and refuse. These fuels and wastes are organic and with complete combustion, the byproducts are CO_2, and water vapor. However, their complete oxidation rarely takes place. In addition, fuels or refuse contain impurities or additives and these are also emitted into the air when burned. Some of these direct products further react within the atmosphere and produce additional indirect products. We may say that air pollution results from gaseous emissions from mainly industry, thermal power stations, automobiles, domestic combustion etc. Following are the important sources of air pollution:

(i) Industrial Emissions

There are a number of industries which are source of air pollution. Petroleum refineries are the major source of gaseous pollutants. The chief gases are SO_2 and NOx. The fertilizer industries release oxides of nitrogen and dust particles of microscopic size. Dust particles may be evolved from the process of drying, burning, calcining, grinding, screening, mixing and packaging. Cement dust is a common air pollutant around cement factories and construction sites which emit plenty of dust, which is potential health hazard. Chemically it is a mixture of oxides of aluminum, silica, potassium, calcium and sodium. Stone crushers and hot mix plants also create a menace. The SPM levels in such areas of stone crushing are more than five times the industrial safety limits. There are many food and fertilizers industries, which emit gaseous pollutants. Many chemical manufacturing industries emit acid vapors in air.

In the past two decades we have seen a tremendous increase in the number of industries in our metros particularly in Delhi, Bombay, Calcutta and Bangalore. Many of these are located in predominantly residential areas. The industries producing sulphuric acid, discharge large quantities of sulfur dioxide in the air and pollute the atmosphere up to a few kilometers. The industries producing fluoride compounds have serious effects on plants, animals and human beings. The toxic effects of fluorides on live stocks arise from ingesting contaminated forage on which fluoride dust has settled. The various iron and steel industries emit sulfur dioxide, oxides of metals and carbon mono and di oxides. Acids are also produced in the atmosphere due to reaction of sulpher di-oxide and nitrogen oxides with water vapors.

(ii) Automobiles

The toxic vehicular exhausts are sources of considerable air pollution, next only to thermal power plants. The ever-increasing vehicular traffic density has posed a continues threat to the ambient air quality. There are over 300 million cars, trucks and buses in the world over and their number increasing rapidly. India, which is likely to have over 30 million vehicles, have more than 65% two-wheelers operating on petrol. In all the major cities of the country about 800 to 1000 tones of pollutants are being emitted into the air daily, of which 50% come from automobiles exhausts. In the major metropolitan cities, vehicular exhaust accounts for 70% of all CO, 50% of all hydrocarbons, 30-40%of all oxides and 30% of all SPM. In cities like Delhi, during peak traffic hours, automobiles of all classes emit as much as 700 Kg of CO, 250 Kg of hydrocarbons and 60 Kg of nitrogen oxides. The two –wheelers and three-wheelers contribute 60% of the total CO, and 83% of total hydrocarbons, whereas heavy traffic vehicles 55 to 80% of the oxides of nitrogen. It is estimated that a car (without cleaning device) on burning 1000 liter of fuel emits 350 Kg CO, 0.6 Kg SO2 ,0.1 Kg lead and 1.5 Kg SPM..

The sources of emission in automobiles are (i) exhaust system, (ii) fuel tank and carburetor and (iii) crankcase. The exhaust produces many air pollutants including unburnt hyrocarbons, CO, NOx lead oxides. There are also traces of aldehydes, esters, ethers, peroxides and ketones which are chemically active and combine to form smog in presence of light. Evaporation from fuel tank goes on constantly due to volatile nature of petrol, causing emission of hydrocarbons. The evaporation through carburetor occurs when engine is stopped and heat builds up, and as much as 12 to 40 ml of fuel is lost during each long stop causing emission of hydrocarbons. Some gas vapor escapes between walls and the piston, which enters the crankcase and then discharges into the atmosphere. This accounts for 25% of the total hydrocarbon emission of an engine.

(iii) Nuclear and Thermal power stations

The catastrophic accident at Chernobyl in USSR in April 1986 is the biggest example of pollution that is being caused by thermal power stations. Radioactive emissions may penetrate through biological tissues in a manner analogous to tiny bullets. They leave no visible marks nor are they felt, but they are capable of breaking molecules within cells. In high doses, radiation may cause enough damage to prevent cell division. In low doses it may damage to DNA molecules. There are a number of thermal power stations and super thermal stations in the country. The National Thermal Power Corporation (NTPC) has setup mammoth coal-powered power stations to augment the energy generation of our country. The coal consumption of thermal plants is several million tones. The chief pollutants are fly ash, SO_2 and other gases hydrocarbons.

EFFECTS OF VARIOUS AIR POLLUTANTS

Some of the effects of various air pollutants have been given as below:

(i) Carbon mono oxide combines with hemoglobin of blood, reducing its O_2 carrying capacity. The gas is fatal over 1000 ppm, causing unconsciousness in an hour and death in four hours. If this gas is inhaled for few hours at even a low concentration of 200 ppm, it causes symptoms of poisoning. Inhaled CO combines with blood hemoglobin to form carboxyhemoglobin about 210 times faster than O_2 does. Formation of carboxyhemoglobin decreases the overall O_2 carrying capacity of blood to cells resulting into oxygen deficiency-hypoxia. At about 200 ppm concentration for 6-8 hours, headache begins, and mental activity gets reduced; above 300 ppm, throbbing headache starts followed by vomiting and collapse; at 500 ppm, man reaches into coma and at 1000 ppm, death occurs.

(ii) The chronic exposures to SO_2 causes intense irritation to eyes and respiratory tract. It is absorbed in the moist passage of upper respiratory tract, leading to swelling and stimulated mucus secretion. Exposure to 1 ppm level of SO_2 causes a constriction of the air passage, and causes significant bronocho-constriction in asthmatics even at a lower (0.05-0.50ppm) concentrations. Moist air and fogs increase the SO_2 dangers due to formation of H_2SO_4 and sulphate ions; H_2SO_4 is a strong irritant (4-20 times) than SO_2.

(iii) Plants are affected at lower concentrations of SO_2 than human beings. Leaf bleaching, necroses and loss of yield are the main damages caused by SO_2. Growth of the ornamental plants Calendula officinalis and Dahlia rosea was adversely affected when exposed to 1 to 2 ppm of SO_2 for two hours.

(iv) SO_2 is also involved in the erosion of building materials as limestone marble, the slate used in roofing, mortar and deterioration of statues. Petroleum refineries, smellers, Kraft mills deteriorate the adjoining historic monuments.

(v) At a low concentration, H_2S causes headache, nausea, collapse, coma and finally death. Unpleasant odor may destroy the appetite at 5 ppm level in some people. A concentration of 250 ppm may cause conjunctivitis and irritation of mucus membranes. Exposure at 500 ppm for 15-30 minutes may cause colic diarrhea and brenchial pneumonia. This gas readily passes through alveolar membrane of the lung and penetrates the blood stream. Death occurs due to respiratory failures.

(vi) The gas nitric oxide is responsible for several photochemical reactions in the atmosphere, particularly in the formation of several secondary pollutants like PAN, O_3, carbonyl compounds etc. in the presence of other organic substances.

(vii) Concentrations of NO_2 in access of 2mg/kg of air causes leaf damage to sensitive plants. In some plants, photosynthetic activity is impaired at a concentration of 0.6 mg/kg. In high concentrations, the oxides of nitrogen cause irritation to the mucus membrane and damage to the respiratory system.

(viii) The important photo chemical products which are produced in the atmosphere as a result of interlinking of oxides of nitrogen, hydrocarbons and ozone in the presence of light include Olefins, aldehydes, ozone, PAN and photochemical smog. Olefins are produced directly form the exhaust and in the atmosphere from ethylene. At very low concentrations of few ppb, they affect plants seriously. They wither the sepals of orchid flowers, retard the opening of carnation flowers and may cause dropping of their petals. At high levels they retard the growth of tomatoes. Aldehydes as HCHO and olefin, acroleins irritate the skin, eyes and upper respiratory tract.

(ix) The particulate matter is injurious to health. Soot, lead particles form exhaust, assents, flyash, volcanic emission, pesticides, H_2SO_4, mist metallic dust, cotton and cement dust etc when inhaled by man cause respiratory diseases such as tuberculosis, and cancer.

In addition to the above there are also many kinds of biological particulate matter that remain suspended in atmosphere. These are bacterial cells, spores, fungal spores, and pollen grains. These cause bronchial, allergy and many other diseases in man, animals and plants.

(x) The various hydrocarbons have carcinogenic effects on lung. They combine with NOx under UV-component of light to form other pollutants like PAN and Ozone (photochemical smog) which cause irritation of eye, nose and throat, and respiratory distress.

CONTROL OF AIR POLLUTION

Some of the methods to control the air pollution are as under:

(i) Removing gaseous pollutants by the process of adsorption. In this process disposal of gaseous pollutants is facilitated by concentrating them on the adsorbents.

(ii) The air pollution is also controlled by using the process of absorption. In this process the gases are passed through absorbers containing liquid absorbents which remove or modify one or more gases which are responsible for the air pollution.

(iii) Facilitating the complete burning of hydrocarbons and organic compounds thus leaving minimum or no unburned compounds which are major causes of air pollution.

(iv) Using low sulpher coal in industries. Sulpher can also be removed from the coal by washing it.

(v) Use of incinerators, settling chambers and electrostatic precipitators in the removal of various pollutants.

(vi) Using proper proportion of gasoline and air in the motor vehicles.

(vii) Using alternative sources of energy such as wind energy, solar energy etc. which do not release any pollutants in the atmosphere.

(viii) Continuous monitoring of air quality will also help in the proper control of air pollution.

(ix) Enforcement of legislation like the Air (Prevention and Control of Pollution) Act, 1981 and The Motor Vehicle Act, 1988 will help in minimizing the release of air pollutants in the air.

(x) By planting trees, using mass transport systems and use of less polluting fuels may also help in the control and reduction of air pollution.

WATER POLLUTION

Water pollution may be defined as "the addition of any substance to water or changing of it's physical and chemical characteristics in any way, which interferes with its use for legitimate purposes". Thus, pollution of natural water implies that it contains a lot of inorganic substances introduced by human activities, which change its quality and are harmful for many living organisms, including man. It is one of the most serious problems in the world, particularly in developing countries.

Normally water is never pure in a chemical sense. It contains impurities of various kinds-dissolved as well as suspended. These include dissolved gases (H_2S, CO_2, NH_3, N_2), dissolved minerals (Ca, Mg, Na salt), suspended matter (clay, silt, sand) and even microbes. These are natural impurities derived from atmosphere, catchments areas and the soil. They are in very low amounts and normally do not pollute water and it is potable. Polluted water, however, is turbid, unpleasant, bad smelling, unfit for drinking, bathing, and washing or other purposes. They are harmful and are vehicles of many diseases as cholera, dysentery, typhoid etc.

CAUSES OF WATER POLLUTION

The chief sources of water pollution are:

1. Oxygen demanding waste

The primary cause of water degradation has been the presence of substances collectively called oxygen demanding wastes. These are primarily organic materials that are oxidized by bacteria to carbon di -oxide and water. Sewage and other oxygen demanding wastes have been classified as water pollutants because their degradation leads to oxygen depletion.

2. Dumping of waste in water bodies

There is uncontrolled dumping of wastes of rural areas, towns and cities into ponds, lakes, stream or rivers. The decomposition of these wastes by aerobic microbes decrease due to higher levels of pollution. The self-purifying ability of the water is lost and water becomes unfit for drinking and other domestic uses. Since decomposition of sewage and other wastes is largely an aerobic process, accumulation of these in water increase its oxygen requirements (BOD).

3. Detergents

Phosphates are the major ingredients of most detergents, They favor the luxuriant growth of algae which form water blooms. This extensive algae growth also consumes most of the available oxygen from water. This decrease in level becomes detrimental to growth of other organisms, which produces a foul smell upon decay.

4. Discharge of untreated sewage

One of the most common primary sources of water pollution is the discharge of untreated or partly treated sewage in water bodies, sometimes due to improper sewage-handling processes of municipal bodies. As oxygen demanding wastes rapidly deplete the DO of water, it is important to estimate the amount of these pollutants in a given body of water. The biochemical oxygen demand (BOD) of water has been a quantity related to the amount of wastes present.

5. Industrial wastes as water pollutants

Many industries, such as steel and paper are situated on the banks of rivers as they require huge amount of water in their manufacturing process. These industries dump their wastes in to rivers. A wide variety of both, inorganic and organic pollutant are present in effluents from breweries, tanneries, dying textiles, paper and pulp mills, steel industries, mining operations etc. The pollutants include oils, greases, plastics, metallic wastes, suspended solids, phenols, toxins, acids, salts, dyes, cyanides, DDT etc., many of which are not readily susceptible to degradation and thus cause serious pollution problems. H2SO4 as acid waste from coal mines is a serious pollutant that increase the hardness of water, has disastrous effect on live organisms and corrodes concrete etc. Na, Cu, Cr, Cd, Hg, Pb, etc. are the heavy metal effluents, discharged from industries.

6. Agricultural discharges

It includes sediments, fertilizers, pesticides and farm animal wastes. All these pollutants can enter waterways as run off from agricultural lands but farm animal wastes are an especially large problem near the large feed lots on which thousands of animals concentrated. Their discharge reach into the water bodies. Artificial fertilizers, pesticides and biocides are the important pollutants which find their ways in to water bodies thus causing water pollution.

GROUND WATER POLLUTION

In most developing countries as ours, most of the underground sources of drinking water, especially in outskirts of larger cities and villages are polluted. For instance trans- Yamuna area of Delhi faces drinking water pollution problem at regular intervals. There had been epidemic of cholera, dysentery and other disease in last couple of years. This is mainly due to inadequate waster supply system in these areas. Ground waster is threatened with pollution from seepage pits, refuse dumps, septic tanks, barnyard manures, transport accidents and different pollutants. Important sources of ground water pollution are sewage and other wastes otherwise. Raw sewage is dumped in shallow soak pits. This gives birth to cholera, hepatitis, dysentery etc., especially in areas with high water table. The industries of woolens, bicycles in areas of Punjab contribute high amount of Ni, Fe, Cu, Cr and cyanides to ground waters.

MARINE POLLUTION

All that what is carried by rivers ultimately ends up in the seas. On their way to sea, rivers receive huge amounts of sewage, garbage, agricultural discharge, biocides, including heavy metals. These all are added to sea. Besides these discharges of oils and petroleum products and dumping of radionuclieds waste into sea also cause marine pollution. Huge quantity of plastic is being added to sea and oceans. Over 50 million lb plastic packing material is being dumped in sea of commercial fleets, whereas over 300 million lb entering through inland waterways in USA. Many marine bird ingest plastic that causes gastro-intestinal eggshgel and tissue damage of egg. Radionucilde waste in sea includes Sr-90, Cs-137, Pu-239, Pu-240.

The pollutants in seas may become dispersed by turbulence and ocean currents or concentrated in the food chain. They may sediment at the bottom by processes like adsorption, precipitation and accumulation. Bioaccumulation in food chain may result into loss of species diversity. The pollution in Baltic sea along the coast of Finland, took place largely from sewage and effluents form bottom fauna. There was seen distinct zonation with extent of pollution. In clear or less polluted water there was rich species diversity which tended to decrease with increasing pollution load. In heavily polluted area, macroscopic benthic animals were absent, but chironomid larvae occurred at the bottom.

In marine water the serious pollutant is oil, particularly when afloat on sea. An spill in oil or petroleum product due to accidents or a deliberate discharge of oil polluted waste brings about pollution. About 285 million gallons of oil are spilled each year into ocean, mostly from transport tankers. This is enough to coat a beach 20 feet wide with half and inch oil layer for 8633 miles. About 50,000 to 250,000 birds are killed every year by oil. The oil is soaked in feathers, displacing the air and thus interferes with buoyancy and maintenance of body temperature, Hydrocarbons and benzpyrene accumulate in food chain and consumption of fish by man may cause cancer. Detergents used to clean up the spill are also harmful to marine life.

Mercury Pollution

Mercury enters water naturally as well as through industrial effluents. It is a potent hazardous substance. Both, inorganic and organic forms are highly poisonous. Methyl mercury gives off vapor. Mercury compounds enter the water body from the effluents and at their bottom these are metabolically converted into methyl mercury compounds by anaerobic microbes. Mercury poisoning is caused due to inactivation of several sulfhydral enzymes by replacement of hydrogen atoms in sulfhydral groups. The antidote, BAL (dimercaprol) is used for mercury poisoning.

Lead Pollution

The chief sources of lead to water are the effluents of lead and lead processing industries.

Fluoride Pollution

Fluoride is also regularly present in water and soil besides air. The crop plants grown in high-fluoride soils in agricultural, non-industrial areas had a fluoride content as high as 300 ppm.

EFFECTS OF WATER POLLUTION

Water pollution has a wide range of effects and these may be gróuped into the following categories:

(i) ***Effects on human health:*** Throughout the world about 70% diseases are water borne or water based. The diseases are either transmitted due to parasites or microorganisms present in the water . Some of the water borne diseases are Typhoid, Cholera, bacterial dysentery, enteritis, infectitious hepatitis, amoebic dysentery, giaridia and schestomiasis. Inadequate water supply for personal hygiene results in skin and eye diseases, examples of such diseases are scabies, leprosy, yaws, trachoma and conjunctivitis.

(ii) ***Decrease photosynthetic rate:*** Water pollution leads decrease in the photosynthetic rate in plants thus reducing the primary productivity. The various sediments responsible for water pollution block the amount of light entering into the system thus not making the light available to the plants.

(iii) ***Decrease the level of dissolved oxygen in the water:*** The oxygen dissolved in the water is used for the

degradation of organic matter which is dumped into water. Thus a lower level of oxygen is made available to the water bodies which is very harmful to aquatic animals and may ultimately lead to their death. The oxygen depletion also leads release of phosphates from the bottom sediments and causes eutrophication of lakes or water bodies.

(iv) Many heavy metals have toxic effects on living beings when they reach to them through polluted water. Mercury readily penetrates the central nervous system of children, lead pollution causes damage to liver and kidney. Higher level of fluoride in water which is used by human beings results in fluorisis.

(v) Pesticides and biocides which find their way to water and pollute have a long ranging effects. These biocides cause considerable harm since their effects are cumulative and lead abnormalities of brain, heart, liver and kidney.

CONTROL OF WATER POLLUTION

(i) Proper laws may be enforced to control the water pollution from point sources. In India we have The Water (Prevention and Control of Pollution) Act, 1974, which is a comprehensive Act in its coverage.

(ii)) Under the National water quality monitoring programme, the water quality monitoring of rivers being done. The programme has been extended and at present, there are 507 monitoring stations in the country spread over all important water bodies. Out of 507 stations, 414 are on rivers, 25 on ground water, 38 on lakes and 30 on canals, creeks, drains and ponds etc.

Taking BOD as indicator of organic pollution, an attempt has been made to estimate the reverine length under different levels of pollution as follows:

High Pollution	BOD > 6 mg/l
Moderate Pollution	BOD= 3-6 mg/l
Relatively Clean	BOD < 3 mg/l

Similarly, the CPCB in collaboration with the Dept. of Ocean Development has identified 173 monitoring station all along the Indian Coast to assess to water quality. Four SPCB have also been involved. Data on 25 parameters are being processed to schemes to control and monitor pollution of the coastal waters.

(iii) An excessive use of chemicals, fertilizers and pesticides should be avoided as these chemicals ultimately find their way to different water bodies and cause severe pollution.

(iv) Proper management of sewage should be carried out by providing a suitable drainage system and over flow of sewage with rain water should be prevented.

(v) Proper water treatment systems should be adopted to treat the polluted water. Water may be treated using any one of the following method: coagulation and flocculation, sedimentation and filtration and disinfection.

SOIL POLLUTION

Soil which is formed by the weathering of rocks, is a very important resource. It provides all the basic needs of all terrestrial organisms in various ways. The major food and fodder for the use of human beings and their animals are grown on soil. However, the formation of soil is affected by a variety of passive and active forces. Various activities of human beings lead pollution of soil not only in urban and industrial areas but also in rural areas. Dumping of various waste generated in various industries and municipalities causes release of various toxic substances which ultimately find their ways in to the soil and ultimately to the ground water. The use of various types of pesticides, herbicides, insecticides, fertilizers etc. to increase the crop production also leads to the severe contamination of soil. In the soil pollution the different pollutants remain in direct contact of soil for a longer period and enter in the food chain of the nature and also in the air and water.

SOURCES OF SOIL POLLUTION

The important sources of soil pollution are as under.

(i) Agriculture practices

The use of large quantity of fertilizers, insecticides, herbicides, weedicides and various soil conditioning agents pollute the soil to a great extent. Similarly excessive amount of plant and animal wastes is also adding to the problem of soil pollution. The fertilizers used to improve the fertility of soil also contaminate it with their impurities. The excessive use of nitrogen, phosphorus and potassium also inhances to this problem.

The increasing use of pesticides to control various pests is causing a stress on the natural environment. During last one decade the number of pesticides has increased manifolds and important among these are organophosphates like malathion, parathion, ethion, trithion etc., and the chlorinated hydrocarbons e.g., BHC, DDT, aldrin, lindane, chlordane and endosulphan. When these pesticides are applied for the contreol of a pest, these are released to the soil and their remanants contaminate to the soil. The herbicides used at the time of seedling remain active for a long time thus add to the soil contamination. Various soil conditioners and fumigants used to increase and protect the soil fertility have several toxic metals which keep on accumulating in the soil

(ii) Urban waste

The large amount of waste being generated in the urban areas contribute to soil pollution. This waste which is normally termed as refuse contains garbage, plastics, metals, fibre, paper, glasses, containers, fuel residues etc. The generation of this waste in huge quantitities has become a problem world over. The capital of India generates approximately 5000 tones of garbage everyday. The soil pollutants ultimately find their way to ground water thus affecting it severely.

(iii) Industrial waste

This is one of the most important soil pollutant and its disposal is becoming a serious problem with every passing day. The industrial pollutants are mainly discharged from sugar factories, oil refineries, tanneries, distilleries, fertilizer and pesticide industries and also from chemical and steel factories. The metal processing industries, drugs, cement, petroleum and coal and mineral industries are other important generators of such waste material. Fly ash, which is mainly being generated by the thermal power plants is another important soil pollutant. All the pollutants released by these industries ultimately affect and alter the biological and chemical properties of the soil.

EFFECTS OF SOIL POLLTTANTS

The various effects of soil pollutants are as under:

(i) The beneficial microorganisms such as bacteria etc. are destroyed by metallic contaminents. These contaminents such as Cr, Zn, Pb, Cd, Hg etc. remain in the soil and their accumulation in the soil for a long period is injurious to the living organisms.

(ii) Industrial wastes from fertilizers, steel, paper and textile industries are extremely toxic are sometimes very toxic to living beings,. These are transceferred to different organisms through the food chain and lead to a variety of harmful effects in living world.

(iii) Solid waste which is directly thrown in the open land results into offensive odour and causes a variety of diseases and ground water pollution.

(iv) Radioactive pollutants dumped into soil produce great human misearies and are responsible for a number of diseases in human beings.

(v) The various volatile materials which are released into the spoil when released into the atmosdphere, contaminate it through the plants they find their way in to human beings causing disruption in the various physiological processes.

(vi) Various fertilizers and pesticides addeded in to the soil are responsible for many ill effects in otrher biota.

(vii) Many pathogenic bacteria which are present in soil cause serious threats to human health.

CONTROL OF SOIL POLLUTION

Soil pollution caused by various sources can be reduced using different methods:

(i) No effluents should be discharged to the soil before these are given proper treatment.

(ii) The various solid waste materials should be collected and as far as possible are recycled.

(iii) Soil pollution due to pesticides may be reduced by using biodegradable materials

(iv) Cattle dung and other biodegradable waste should be used to produce biogas

NOISE POLLUTION

Noise pollution is a direct result of technological development. Noise, with its ever increasing effects on human beings and on the environment, is defined as an acoustic fact which is unpleasant and arouses disturbing feeling or as the totality of unwanted, undesired sounds. The human ear is constantly being assailed by man made sound from all sides, and there remain few places in population areas where relative quiet prevails. What do airplanes, trains, cars, and numatic drills, and radio and television sets have in common? They all produce noise, the most dangerous pollutant of man's environment. Noise has become a permanent part of our lives these days because of the development of machinery, industry and technology. Noise harms the body and mind. Noise not only cause irritation or annoyance but it constricts your arteries, increases the flow of adrenaline and forces your heart to work faster.

The word noise (Latin nausea) is usually defined as unwanted or unpleasant sound that causes discomfort. Noise is also defined as "wrong sound, in the wrong place at the wrong time". Noise pollution means, "the unwanted sound dumped into the atmosphere leading to health hazards"

Even if noise is not sufficiently loud to constitute direct threat to human health, continuous exposure to noise shortens human beings sleeping hours and reduces his productivity. The decrease in productivity affects both city dwellers and people who work in factories or in the industrial areas. In China, till the third century BC, noise was used as a method of torture instead of hanging men for dangerous crime. The importance of noise as a pollutant having a deleterious effect on peace of mind and beauty of environment is increasing every day. Formerly noise was limited only to the industry. This too was not much as there were only few industries. These days there has been rapid industrial growth. Moreover, there has been population explosion, due to which there is heavy traffic, urban crowd and electric equipment (luxury items and entertainment). All these have added to the noise nuisance in environment. In our country, beside these the two other factors are the religious and social functions, which increase the gravity of situation. The intensity of noise from some sources is given as below:

Sources	**Intensity (dB)**
Breathing	10
Broadcasting studio	20
Soft whisper	20-30
Trickling clock	30
Library	30-35
Low volume radio	35-40
Normal conversation	35-60
Telephone	60
Office noise	60-80
Alarm clock	70-80
Traffic	50-90
Trunk	90
Motor cycle	105
Lion's roar (12')	105-110
Jet fly (over 1000')	100-110
Train whistle (50')	110
Air craft (100') (Prokeller driven)	110-120
Pneumatic drill	110-120
Commercial jet (air craft (100'))	120-140
Jet take off (300')	120
Space rocket (launching)	170-180

SOURCES AND EFFECTS OF NOISE

Sources of noise are numerous but these may be broadly classified in to two classes: i. Industrial and (ii) Non-industrial

The main contributors to noise are factories and industries, transportation (air, rail and road). The disturbing qualities of noise emitted by industrial premises are generally its loudness, its distinguishing feature such as tonal or impulsive components and its intermitancy and duration.

Among the non-industrial sources, important ones are as follows (a) Loudspeakers, (b) automobiles (c) aircrafts (d) trains (e) construction works (f) radio, microphone etc. The chief man-made sources in urban areas are automobiles, factories, industries, trains, airplanes. Noise makers are horns, sirens, lawn mover, musical instruments, TV, radio, transistor, telephone, dog, loudspeaker, washing machines, vacuum cleaner, food mixers, pressure cookers, fans, air conditioners, coolers. Ever since the industrial revolution, there has been doubling of environmental noise in every 10 years.

Effects: Noise has been found to interfere with our activities at three levels : (I) audiological level (ii) biological level, interfering with the biological functioning of the body and (c) behavioral level, affecting the sociological behavior of the subjects.

1. Auditory effects : These include auditory fatigue, and deafness. Auditory fatigue appearing the 90 dB and may be associated with side effects as whistling and buzzing in ears. Deafness can be caused due to continuous noise exposure. Temporary deafness occurs at 4000-6000 hz. Permanent loss of hearing occurs at 100 dB.

2. Non-auditory effects : These are (i) interference with speech communication, (ii) annoyance, (iii) loss of working and (iv) physiological disorders

   (a) Interference with speech communication : A noise of 50-60 dB commonly interferes with speech; sound of warning (signal) may be misunderstood.

   (b) Annoyance : Balanced person express great annoyance at even low level of noise as crowd, highway, radio etc. The effects are ill temper, bricking.

(c) Loss in working efficiency : There develop tiredness and those doing mental work may put to deterioration in their efficiency or even complete loss of ability to work.

(d) Physiological disorders : A number of physiological disorders develop due to imbalance in functioning of the body. These are neurosis, anxiety, insomnia, hypertension, hepatic diseases, behavioural and emotional stress, increase in sweating, giddiness, nausea fatigue etc. Noise also cause visual disturbance, and reduces depth and quality of sleep thus affecting overall mental and physical health. Other effects are, undesirable changes in respiration, circulation of blood in skin and gastrointestinal activity. Noise pollution also causes incidence of peptic ulcers.

Continuous noise causes an increase in cholesterol level resulting in the constriction of blood vessels making you prone to heart attack and strokes. There may be still births and usually low weight children born to mothers living near airports.

Supersonic air planes create a shock wave called sonic boon, which produces a startle effect that can be more harmful than a continuous noise. The sonic boon may spread in an area of 10 to 80 miles and when its hits the ground it damages window pans and building structures. This may also fasten the human fetus heart rate. Some of the important health hazards of noise are as follows:

Noise intensity (dB)	Health hazards
80	Annoyance
90	Hearing damage
95	Very annoying
110	Stimulation of reception in skin
120	Pain threshold

130-135	Nausea, vomiting, dizziness
140	Pain in ear
150	Burning of skin
160	Rupture of tympanic membrane
180	Major permanent damage in short time

CONTROL OF NOISE POLLUTION

(i) **Control at source:** This can be done by (i) designing and fabricating silencing devices in aircraft engines, automobiles industrial machines and home appliance and (ii) segregating the noisy machines. The proper design of equipment to minimize the noise generation has been somewhat a complex engineering problem needing a strong background in the fundamentals of vibration, fluid mechanics, dynamics, materials and machine design. It is also important to operate all the equipment at the design conditions. Operating equipment at designed pressure and speed should result in minimum noise generation. It should be apparent that maintenance is an important element in controlling the noise generated by equipment.

(ii) **Transmission control :** This can be achieved by covering the room walls with sound absorber as acoustic tiles and construction of enclosures around industrial machinery.

(iii) **To protect exposed person :** The worker exposed to noise can be provided with wearing devices as ear plugs and ear muffs.

(iv) **To create vegetation cover :** Plants absorb and dissipate sound energy and thus act as buffer zone. Trees should be planted along high ways, streets and other places. Ashok, Neem, Tamarind etc. are good for this purpose.

(v) **Control through law:** Silence zones must be created near schools, hospitals and indiscriminate use of loudspeakers at public places must be prevented by laws. Adequate restriction must be put on unnecessary use of horns and vehicles plying with out silencers. There are already such laws in some countries as UK and USA. In India, we have Motors vehicles Act which provides restriction on trucks using double sirens while passing through some localities. But this is not enough. In Delhi and Bombay, there are flights round the clock at airport. Delhi is closely following Bombay in noise pollution and if adequate steps are not taken to reduce sound level, more than 50% of Delhi will be affected in coming years. Restriction may be put on air craft flight at mid-night.

(vi) **Education :** No doubt we have the technology to control nearly every kind of noise. We know very well about the sources and impact of noise pollution. But the main problem is that the people are not aware that the noise is one of the main environmental problems. Public must be made aware and educated about noise nuisance through adequate news media, lectures and other programs. The movement against noise pollution is very weak in India. The main reasons being that most of us do not consider noise as a pollution but as a part of routine life.

THERMAL POLLUTION

One of the most significant non-chemical ways in which man impacts aquatic systems is through artificially heating up the water of the receiving system. Water has a much greater heat capacity than air and therefore undergoes less temperature fluctuations than the terrestrial environment. So, the aquatic organisms experience less temperature fluctuation and therefore have evolved to be much less suited to temperature changes with above in mind thermal pollution can be defined as "The warming of air aquatic ecosystem to the point where desirable organisms are adversely affected". Thermal pollution takes place because many electric-generating companies use water in the process of coaling their generators. This heated water is then released into the water body from which it was taken, causing a warming of the surface water.

EFFECTS OF THERMAL POLLUTION

(i) Increasing the water temperature of a system is harmful since it generally alters the physiochemical & biological characteristics of that system. In addition to the possibility of decreasing or eliminating various aquatic forms, it may also stimulate spawning at a time of year when food supplies are restricted. This leads to starvation of the newly spawned individuals in the population.

(ii) High temperature also decrease the density and viscosity of water, causing an increased settling rate of suspended solids. Evaporation rate is incrased.

(iii) The addition of heated water to a cooler, water body may accelerate the lowering of DO levels due to density differences between two. The less dense warm water tend to form a layer on top of the cooler, more dense water. This occurs particularly when the body of cool receiving water is deep. The resulting envelop of hot water cannot dissolve as much atmospheric oxygen ad the underlying cold water, which is denied contact with atmosphere.

(iv) Another effect of this stratification may show up downstream from a dam when the oxygen deficient levels get discharged through the lower gates of a dam. Serious effects on down stream aquatic life may result.

(v) Both the withdrawal as well as the discharge of water foe cooling can have adverse environmental impact. The withdrawal of large volumes of water for cooling can result in some organisms being impinged on intake screens. They also may take a severe battering by the displacement of the large volumes of water. Smaller creatures including plankton, fish eggs and small fish are sent past the screens and are either killed by chemical treatment or directly by the heat itself.

(vi) Elevated temperatures can affect" temperature cues" governing the reproductive cycle of many species. Fish often spawn by temperature cues to give birth when food supplies for young will be available. If a false temperature cue from thermal pollution initiated spawning, the necessary food supplies may not be available when needed. Elevated temperatures also can cause fish eggs to hatch earlier. Trout eggs, for example, hatch in 165 days when incubated at 37°F. At 54°, they in only 32 days.

(vii) Warmer water is lighter and generally floats on colder denser water and can affect surface-dwelling species directly or act as a barrier. It can affect the food supply, which may not be available for a consumer. Because temperature can effect solubility of substances, some species such as corals are limited to warm waters where the solubility of calcium carbonate is lower.

(viii) Because elevated temperature will cause water to hold less oxygen and also will cause respiration to double for each 10°C rise, the potential effect is obvious. This is particularly for predatory fish.

CONTROL

The problem of thermal pollution can be managed by using artificial cooling lakes, spray ponds and cooling towers. Cooling ponds are man made bodies of water where heated effluents are allowed to discharge and water for cooling purposes is drawn from beneath the surface at the other end of lake. In spray ponds water is received from condensers and sprayed through nozzles where fine droplets of water are formed. The heat of water is dissipated into the atmosphere from these small droplets. Cooling towers are able to transfer heat from water to the atmosphere generally through evaporation of water. The cooling towers may be either wet cooling towers or dry cooling towers.

NUCLEAR HAZARDS

Man-made source of radiation are of various nature, ranging from a household article to atomic explosions. A number of articles, which we use in our daily life, such as luminous clocks, watches, television sets and even telephone dials, all contribute to the radiation to some extent which may exceed by 1% of the total background radiation received by a person. Besides, people are significantly exposed to radiations received from medical use of x-rays and radioactive isotopes. A small amount of radioactivity is also introduced into the environment by the use of radionuclides in industry, medicine and research.

The nuclear explosions carried out for bumb testing or for beneficial uses such as for mining and excavation of harbours and canals, release huge quantity of radioactive elements along with heat open in the environment. These radioisotopes fused with dust and dirt rise in the atmosphere along with the hot gases and then slowly fall onto various surfaces on the earth by gravity. This is called 'radioactive fall-out'. The radionuclide in fall-out differ from other radioactive wastes in that they remain fused with iron, silica and dust and become insoluble. They look like particles of marble with different colours with the size variable from few microns to as large as snow particles. The radioactive elements can originate after the explosion by direct fission of the fuel or by absorption of released neutrons by ambient elements and molecules. Examples of the common radioisotopes originated directly are strontium-90 and cesium- 137 and of induced radioisotopes are cobalt-60, iron-59, zinc-65 and manganese-54. In aquatic environments, several fall-out elements such 60Co, 59Fe, 65Zn, 54Mn, 144Ce, 95Zr and 106Rn form complexes with organic detritus, from where they are readily transferred to the aquatic organisms.

The movement of radioactive materials in the environment and their ultimate effects on Man and illustrated in Fig. The radioactive pollutants reaching the freshwater resources or oceans are rapidly lost to the sediments (Hethrington, 1976; Emery and Clopfer, 1976) and may bioaccumulate in the body of plant and animals directly or through the food chains (biomagnification). In the body of the organisms, they behave chemically in the same manner as their stable counterparts, but are more dangerous because of the radiation which they emit internally. Foster and Mc Common (1972) have demonstrated the accumulation of phosphorus –32 and zinc 65 in substantially higher quantities in fish. Algae and macrophytes also concentrate the radio nuclides in greater amounts from ambient water. The accumulation power of organisms for radio nuclides can be demonstrated by the concentration factors (CF), i.e. the ratio of the element in organisms to that in water. The CF as high as 20,00,000 has been reported for phosphorus-32 in eggs of certain ducks. The concentration factors reported for strontium-90 in aquatic life are:

Filamentous algae	5,00,000
Phytoplankton	75,000
Insect larvae	1,00,000
Fish	30,000 – 70,000
Bone of water birds	500

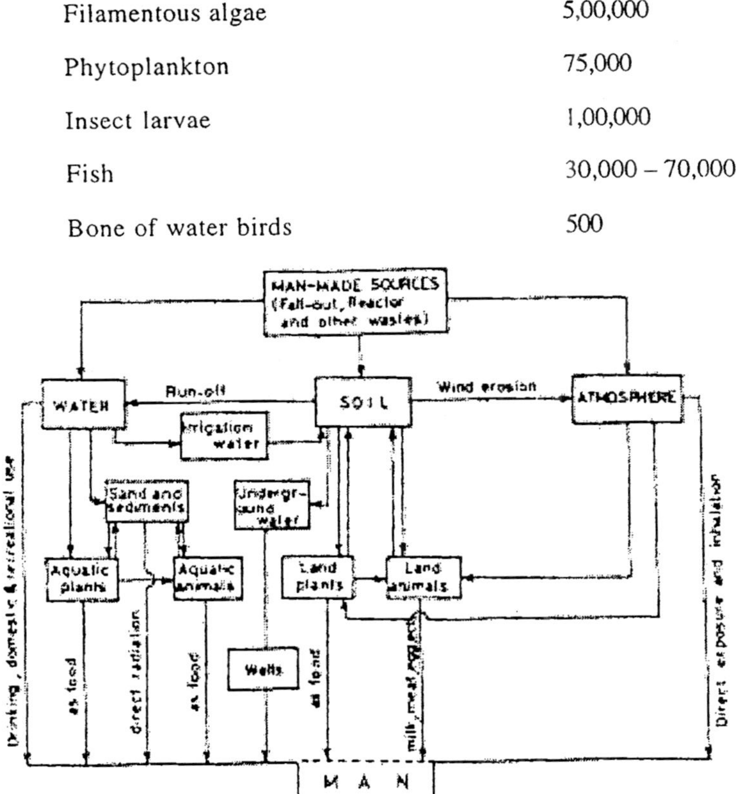

Movement of radioactive materials in the environment affecting the Man

Man is the ultimate sufferer in this manner after consuming the aquatic food. Man may also be exposed to radiation while using the contaminated water for his various needs such as for drinking, cooking, washing and bathing, however, such a dose shall be extremely low, but long term cumulative dose may be of concern in some cases. The irrigation by contaminated water will pollute the soil from where the radionuclides are transferred to the crops. Soil also gets polluted by a direct release of low-activity waste waters and by radioactive fall-out. The radioactivity from the soil moves through the food chains and reaches the Man after consumption of crops, meat, milk and eggs etc. The underground waters may also receive radionuclides after their leaching from the soil. The radioactivity released into the atmosphere is rapidly diluted by atmospheric processes, but it may be of concern to Man in certain highly contaminated areas, for example, in the vicinity of atomic explosions or in atomic power plants. The atmospheric fall-out depositing directly onto the leaves is efficiently passed on to the grazing animals, such as cattle and reach to us. Cesium-137 and strontium-90 are two most important radionuclides found to reach humans in this manner. In aquatic systems, fish eggs have been found to be highly sensitive to even low levels of radioactivity (Ophel, 1976). Fish species, those with low fecundity are greatly affected the radioactivity present in water. Fish and other populations are also affected by the radioactivity present in water. Fish and other populations are also affected by mutations leading to abnormal organisms. In general, according to Ophel (1976), fish are the most sensitive organisms in water to irradiation followed by crustaceans, mollusks, algae (except Cyanophyceae) and protozoa. Human responses to irradiation are greatly dependent upon the level of the dose given and the time of exposure and range from simple sublethal effects to death in acute cases. Some important effects, produces by different radiation doses.

Radiation Dose (Rads)	Human responses
650 or above	Death in few hours or in days.
400	30 days LD50 (Death of 50% humans within 30 days)
100-250	Sublethal dose, causes nausea, vomiting and diarrhoea within hours, itching, burning and ulceration of skin, loss of hair, hemorrhages just below skin, decline of red and white blood corpuscles and loss of ability to fight diseases, genetic, mutations.
Less than 100	Delayed effects such as cancer, leukemia, sterility cataracts and reduction of life span, genetic mutation.

CONTROL OF RADIOACTIVE POLLUTION

Out of all the sources, only artificial radioactivity is in the scope of human intervention where control can be made. However, little can be done when the radioactivity is already released into the environment from the source. Similarly the control of natural radioactivity is also not possible. The radioactive waste concerned with water pollution are usually in the liquid or solid state. The various kinds of waste pose different problems for disposal; techniques suitable for one kind may be useless or risky for another. All techniques of course have a single goal to ensure that radioactive constituents of the wastes are not released into the environment causing any harm to organisms and in particular to Man. All low or high-level wastes have a tremendous capacity to pollute the environment. As low level wastes are often produces in large quantities, their containment is not possible. They are usually subjected to a treatment for removal of radioactivity and then discharged in the bodies of water or on land in usual way. High level wastes, on the other hand, cannot be disposed off free in the environment, but have to be concentrated, contained and stored out of the Man's environment. Some important methods of treatment and disposal of radioactive wastes are given below :

(a) Precipitation

(b) Ion Exchange

(c) Hydrofracture

(d) Bituminization

SOLID WASTE MANAGEMENT

It has already been pointed out that waste is a resource at the wrong place, time and in concentration (Das, 1992). Thus a waste can turn out to be a valuable materials. The solid waste is generated from the natural as well as anthropogenic systems. Amongst the waste generated through natural processes are the solid wastes of the forests, running water etc. Anthropogenic solid wastes originates from basic three types of human activities:

(i) **Agricultural solid wastes :** The solid wastes generated as a result of agricultural activities fall under this category. These consists of crop wastes, cattle dung etc.

(ii) **Urban Solid Wastes :** The solid wastes of urban habitats consist of organic and inorganic materials generated from the household, commercial, institutional, public amenities, dairies and small scale industries operating the city, garages, workshops etc. The solid wastes in the cidities cause nuisance An average Indian produces about 0.4 kg solid waste per day. More than 500 samples collected from different parts of the country gave following composition of wastes:

Paper/cardboard	5.25%	-	10.0%
Plastic items	0.6	-	0.9%
Glass, ceramics, pebbles	0.1	-	0.7%
Dust/sand	30	-	40%
Metallic refuses	0.6	-	1.0%
Vegetables, organic matter, foodstuffs	50	-	65%

Presently the solid waste management technologies are not being fully utilized and much attention is needed for the purpose. There is insufficient system of collection and processing. The organic content start decomposing and cause obnoxious smell. These also result in the spread of certain disease like gastroenteritis, dysentery, jaundice, cholera, malaria etc. In order to meet the challenge of municipal solid waste management there is need to develop a cost effective sustainable technology to sustain life in the urban habitats. The technology has thus to be developed through which the waste can be converted into useful material. This will become economically productive besides cleaning the environment. Final disposal of municipal solid wastes can be carried out by methods like sanitary land filling, incineration, composting etc.

Sanitary land filling

The process of sanitary land filling is one of the oldest methods of the disposal of municipal solid wastes. In this process the waste is deposited in low lying areas and in this way land is reclaimed for habitation. This method is still the most commonly used method for the disposal of solid wastes of the city.

Incineration

Incineration is a combustion process to burn unwanted material into gases and residual simple solids. It also generates toxic chemicals- dioxin and furan. The process is now blamed for releasing certain toxic chemicals in the atmosphere. The emissions consists of oxides of nitrogen, sulphur, carbon together with particulate mater. It also contain chlorinated organic compounds like chlorophenol, furan chlorobenzene etc. The incinerators cannot be taken to be free of hazards de to toxic emissions in form of gases as well as in bottom and fly ash. The attempts are being made to design an eco-friendly incinerator.

Composting

The disposal of organic matter can be accomplished through the process of composting. This can be of two types depending upon the supply of air as aerobic and anaerobic. Both these processes are carried out by micro-organisms. Organic matter is separated from the garbage and are subjected to microbial activity ion digesters which may either be aerobic. Aerobic digestion leads to the formation of black organic manure and carbon dioxide. The anaerobic digestion results in the formation of methane and organic manure. Methane can be used as a fuel.

An alternate method of biodegradation of organic matter involves the use of earthworms to convert organic matter into manure. This technology is known as vermiculture biotechnology. In this process earthworms are allowed to grow on organic waste by providing adequate moist condition and the earthworms convert organic matter in compost.

Hospital Solid Wastes

Hospital solid wastes have to be tackled very carefully as these can cause severe health problems for the community. Different types of matter be separated at source. The organic matter consisting of human body wastes; blood and other body fluid soaked cotton, cloth etc. microbiological wastes; highly infectious wastes have to be incinerated in especially designed incinerators for the purpose. Glasswares, plastics and metal parts be separated and sent for recycling.

Industrial Solid Wastes

In municipal solid waste all varieties of wastes remain mixed, hence it requires their separation before they can be treated separately. The industrial solid waste, however, remain separate as they are generated at different stages of production. Government of India made a Technology Policy Statement, 1993 advocating proper waste management activities in the country. One of the objectives of the policy is for the recycling of the waste material and make full utilization of bye products and ensure harmony with the environment, preserve the ecological balance and improve the quality of habitat. The main ingredients of solid waste are residual ash in the furnace and fly ash coming out of chimney. Usually these are dumped on land and damage the habitat. Techniques have now been developed in which furnace slag after being ground are mixed with cement. The fly ash is being utilized to make bricks for building purpose.

To meet the challenges of increasing solid waste materials, it has been proposed to follow 3 Rs i.e. Reduce, Reuse and Recycle. For reduction of solid waste appropriate technology has to be opted. These are known as cleaner or environment friendly technologies. The waste reduction can also be achieved by reducing consumerism to the extent of its bare need for the sustenance. This concept once inculcated in the society the quantity of waste generated shall drastically be reduced. Recycling amounts to conservation of resources as a product at the end of its service life is transformed into another useful product. In the process of recycling a technology has to be adopted in which the recyclable materials are separated out and sent to proper place where it is transformed into certain other useful materials. The advantage of this process lies in the fact that valuable returns are obtained from the garbage. In other words it amounts the generation of wealth from the waste. Above table provides details of the useful materials produces from some agricultural, urban and industrial solid wastes indicating how can be generated from the wastes.

Table : Showing some of the waste put to useful materials

S.N.	Waste material	Useful material
1	2	3
1.	Bagasse	Coal pellets made with wheat straw
2.	Banana leaves	Fuel
3.	Biomass of plant and animal origin	Compost Biogas methane (Fuel) Charcoal Liquid fuel Protein Amino acids
4.	Chrome sludge	Coloured brick
5.	Corn	Packing material
6.	Fly ash	Bricks Cement additive Catalyst Timber Paint Glass fibres Road making
7.	Garbage/sewage	Fuel pellets, Methane Electricity Biofertilizer
8.	Grains	Chemicals
9.	Jute waste	Paper Board
10.	Maize cobs	Furan with bagasse Sugars
11.	Old tyres	Microbial degradation Rubber

1	2	3
12.	Paper waste	Recycled paper
13.	Plastic waste	Fuel Oil Floor tiles Decorative housing material Heat Electricity Pencil Detergent based stabilizer molecule
14.	Potatoes	Paint
15.	Silk industry waste	Poultry feed
16	Wood	Fuel gas

ROLE OF INDIVIDUALS IN PREVENTION OF POLLUTION

The role of an individual in maintaining a pollution free, pure and congenial environment and in preserving its resources is actually the need of the hour. Individuals can, however, play an important role in abatement of air, water, soil or noise pollution in the following simple manners:

(i) Use less harmful substances instead of commercial chemicals for most household cleaners.

(ii) Use low-phosphate, phosphate-free or biodegradable dishwashing liquid, laundry detergent, and shampoo.

(iii) Don't use water fresheners in toilets .

(iv) Use manure or compost instead of commercial inorganic fertilizers to fertilize gardens and yard plant.

(v) Use biological methods or integrated pest management to control garden, yard, and household pests.

(vi) Don't pour pesticides, paints, solvents, oils, or other products containing harmful chemicals down drain or on the ground. Contact the authorities responsible for their disposal.

(vii) Recycle old motor oil and antifreeze at an auto service center that has an oil recycling program.

(viii) If you get water from a private well or suspect that municipal water is contaminated, have tested by an EPA certified laboratory for lead, nitrates, trihalomethanes, radon, volatile, organic compounds and pesticides.

(ix) Run water from taps for several minutes every morning before using the water for drinking or cooking. Save it and use it to water plants.

(x) If you have a septic tank, monitor it yearly and have it cleaned out every three to five years by a reputable contractor so that it won't contribute to groundwater pollution. Do not use a septic tank cleaner, which contain toxic chemicals that can kill bacteria important to sewage decomposition and that can contaminate groundwater if systems malfunctions.

(xi) Support ecological land-use planning in your community.

(xii) Get to know your local water bodies and form watch-dog groups to help monitor, protect, and restore them.

(xiii) Support efforts to clean up river fronts and harbors.

(xiv) Support tougher water and air pollution control laws and their enforcement at the local, state, and national levels with emphasis on pollution prevention.

(xv) Test for radom and taking corrective measures as needed.

(xvi) Install air-to-air heat exchangers or regularly ventilate your house by opening windows.

(xvii) Don't buy synthetic wall-to-wall carpeting, furniture, and other products containing formaldehyde, use "low-emitting formaldehyde" or non formaldehyde building materials.

(xviii) Reduce indoor levels of formaldehyde and several other toxic gases by using houseplants. About 20 plants can help clean the air in a typical home. Plants should be potted with mixture of soil and granular charcoal (which absorbs organic air pollutants).

(xix) Change air filters regularly, clean air conditioning systems, and empty dehumifider water trays frequently.

(xx) Don't store gasoline, solvents, or other volatile hazardous chemicals inside a home or attached garage.

(xxi) Don't use commercial room deodorizers or air fresheners.

(xxii) Don't use aerosol spray products.

(xxiii) Don't smoke. If you must smoke, do it outside or in closed room vented to it outside.

(xxiv) Have people take off their shoes when entering your room this greatly reduces indoor levels of toxic lead dust and pesticides picked up by the shoe bottoms, which transfer then to floors and especially to indoor carpets.

(xxv) Make sure that wood-burning stoves, fireplaces, and kerosene-and gas-burning heaters are properly installed, vented, and maintained.

POLLUTION CASE STUDIES

Choking of South Asia by a Blanket of Smog

According to a report published in Times of India dated 13th August, 2004, India and rest of the South Asia is covered by a deadly, three km. deep blanket of pollution, which is readily changing monsoon patterns, causing draughts and literally killing hundreds of thousands of people by respiratory diseases. A group of scientists working with the United Nations Environment Program on the Asian Brown Haze has released this study. It said that the vast pollution parcel could endanger the economic success of South Asian Countries, particularly India. The haze has been identified as a deadly cocktail of ash, acids, aerosols and other particles. The study also says that the worst may be yet to come and the effects of the haze would intensify over the next 30 years. Already, higher levels of respiratory diseases are leading to several hundreds of thousands of premature deaths, as revealed by data from seven Indian cities, including Ahmedabad, Kolkatta, Delhi and Mumbai. The Indian data suggested some kind of air pollution was responsible for 24,000 annual premature deaths in the early 1990s, said the report. A few year later, the number rose to an estimated 37,000 per year.

UNEP executive director, Klaus Toepfer says, "In India they are expecting more than two million people to die because of the incomplete burning of biomass." The open fires used by the majority of ordinary Indians to cook their food also contribute to this. This pollution parcel, which stretches three kilometers high, can travel half way round the globe in a week, thus having global implications

The Bhopal Episode

During the night of 2nd December 1984 one of the world's worst industrial accident took place in the capital of Madhya Pradesh. The accident took place at Union Carbide Company, which used to manufacture a pesticide named as Carbaryl for which methyl iso-cynide(MIC) was being used. The accident was caused by the introduction of water into MIC holding tank. The reactions thus initiated generasted large volumes of the gas, forcing the emergency release of pressure. The gas escaped while the chemical scrubbers which would have treated the gas were not working. Other safety devices also did not work. Later on it was also revealed that many safety procedures were bypassed and the standard of operation in Indian plant did not match those at other plants of Union carbide. Duringthe accident about 40 tons of MIC leaked into the atmosphere. This gas at lower concentration affects the lungs and eyes and causes irritation in the skin. Higher amounts remove oxygen from the lungs and cause death. The gas spread over an area of 40 square km and about 4000 people were killed and more than 1,50,000 were injured. The majority of deaths were related to pulmonary oedemas, however, the other wide ranging ailments were also caused by the gas. In an out of court settlement reached in February 1989, the company agreed to pay US dollar 470 million for the damage it caused during the disaster. However, the truth of the entire episode is that the sufferers did not get the proper relief and compensation. And the effects are so wide ranging even the offsprings of the injured ones are also facing the health problems.

Higher Concentration of SPM in Air

According to a reprt of. Centre for Science and Environment (2006) nearly 52,000 people died prematurely in 36 Indian cities due to high level of suspended particulate matter in 1995. It was 28% higher than 40,000 people in 1991-92. The number of people requiring medical assistance due to high levels of SPM in these cities also increased from 19 million in 1991-92 to 25 million in 1995. The average annual suspended particulate matter in the ambient air of some of the cities in India during 1995 was as under:

Cities	SPM (mg per cubic meter)	CBCB Rating
Patna	330.7	Critical
Faridabad	318.3	Critical
Bhilai	288.9	Critical
Jalandhar	287.5	Critical
Dhanbad	275.0	Critical
Surat	267.C	Critical
Satna	254.5	Critical
Kota	241.8	Critical
Nagpur	186.2	High
Kottayam	146.0	High
Jamshedpur	118.0	Moderate
Kochi	113.3	Moderate
Haldia	103.2	Moderate
Guwahati	93.3	Moderate
Shillong	52.5	Low

(*Source:* Down to earth, 2006, Health and Env. In India)

CHAPTER-XII

Diaster Management

The term disaster owes its origin to the French word 'Destre' which is combination of two terms 'des' meaning bad or evil and 'astre' meaning star, when combined together the expression is bad or evil star. In earlier days a disaster was considered to be the loss due to some unfavorable star. This concept has now changed and the term disaster is commonly used to denote any odd event, be natural or man made which brings about sudden and immense miseries. The term disaster has been defined in Oxford dictionary as ***"An event, man made or natural, sudden or progressive, the impact of which is that the affected community must respond through exceptional measures"***. According to the Webster's dictionary a disaster is "A sudden calamitous event producing great material damage, loss and disasters".

Disaster has been defined with following features:

1. Disruption to normal pattern of life. Such disruption is usually severe and may also be sudden, unexpected and widespread.
2. Human beings are affected with loss of life, livelihood, and property, injury, hardship and adverse effect on health.
3. Effects on social structure such as destruction of or damage to infrastructure, buildings, communications and other essential services.
4. Community needs such as shelter, food, clothing, medical assistance and social care.

Thus a complete definition of disaster may be, ***"An event, concentrated in time and space, which threatens a society or a relatively self sufficient subdivision of society with unwanted consequences as a result of the collapse of precautions which has hitherto been culturally accepted as adequate"*** (Turner, 1976). During the last 30 years, natural disasters have caused death of at least three million people and have affected many more millions.

Very often, thé term Hazard and Disaster are used interchangeably, but their connotation and consequences are distinct and must be kept in view. A hazard is the occurrence of a physical phenomenon at a given site that is capable of causing loss and damages. If a hazard such as the cyclone hits an un-populated coastal area, it should not be considered as a disaster. However, it will be called a disaster, if it hits a populated region and life and property in the area gets severely affected.. A hazard, therefore, is a pre-disaster situation that could turn into a full blown disaster if it results in casualties, be it human, socio-economic or infrastructure. It is always the unmanaged hazard that leads to the disastrous results.

Disasters vary in terms of their intensity, disruptive potential, frequency, predictability and duration. Disasters can be cataclysmic or even slow in onset. These occur where vulnerable people are overwhelmed by extreme events or hazards, either natural, human or a combination of both. It is indeed a difficult task to establish a distinction between a natural and man made disaster, though certain mishaps such as Bhopal gas tragedy in 1984, outbreak of plague in Surat, in 1994 and Dropsy epidemic in 1998 clearly fall under the man made category. The last decade of 20th century has witnessed some of the worst catastrophes hitting almost all parts of the country. The devastating earthquake in Uttarkashi disrict of Uttaranchal in October, 1991, in Latur district of Maharashtra in September, 1993 and the most recent and the most destructive earthquake of Gujrat in January, 2001 are still afresh in the memory . Similarly the land slides at Malpa, Rudrapryag and Pithoragarh and the severe floods in different parts of the country are certainly a reflection of the growing frequency and intensity of disasters in our country.

Table. 12.1. Annual average number of people affected by disaster between 1987-1996: Top 10 countries of the world.

Country Affected (million)	Annual average population (million)	Total population (1996)	%Population affected
China	99.07	1232.08	8.04
India	56.56	944.58	5.99
Bangladesh	18.57	120.07	15.47
Ethiopia	4.02	58.24	6.90
Philippines	3.69	69.28	5.33
Australia	2.28	18.05	12.63
Thailand	1.67	58.70	2.84
Sudan	1.48	27.29	5.42
Malawi	1.44	9.84	14.63
Pakistan	1.40	139.97	1.00

(*Source:* World disaster report, 1998)

Among the top 10 countries in terms of average annual loss of life due to natural disasters, except Australia, all are developing countries.

The disasters are of two types. The first category is of natural disasters which are the results of natural phenomenon. Natural disasters are basically related to the following three factors:

1. Wind: Storm, cyclone, hurricane, storm, surge and tidal waves.
2. Water: Flood, cloud burst, flash flood, excessive rain and drought.
3. Earth related: Earthquake, tsunamis, avalanches, landslides and volcanic eruption.

But modern technology is capable of predicting some of these natural calamities. However, warnings issued before such an event are often ignored.

The second types of disasters are of anthropogenic origin. The man made disasters are as under:

1. War/ battle/ hostile enemy actions.
2. Arson/ sabotage/ internal disturbance/ riots.
3. Accidents of vehicles/ trains/ aircraft/ ships.
4. Industrial accidents/ explosion of boilers/ gas cylinders or gas chambers/ gas leaks.
5. Fire and forest fires.
6. Nuclear explosion/ radio active leakage.
7. Ecological disasters like deforestation/ soil erosion/ air/ water pollution.

FLOODS

Floods are of two types, river floods and coastal floods. A river flood occurs when a high river flow exceeds the capacity of the channel and excess water flows out on to the flood plain. There are four main causes of river floods:

a. Heavy precipitation can result in river floods when extremely intense rainfall over a drainage basin gives more that a flood is generated. Alternatively, prolonged and continuous precipitation can be the cause of flooding, for example when successive days have rainstorms for periods of several weeks so tat the ground is saturated, infiltration is not possible, and large amounts of surface runoff lead to the generation of a river flood.

b. Floods can also be generated by snow melt. This occurs annually in some areas of the world, such as Arctic and mountain areas that are covered in snow during the winter so that the spring or melt season gives high discharges as the snow melts. Snow-melt floods can also occur occasionally in other areas if occasional deep snowfalls melt rapidly, often associated with rainfall, and produce large amounts of water leading to river floods.

c. The sudden release of water from ice-dammed lakes can induce river floods, In Iceland these are given the special name of Jokulhlaups, which are flood waves that roar down the valley when the water from an ice-dammed lake finds an outlet. Less dramatic floods can occur along rivers where natural dams of debris, such as dams of trees along Candian rivers, lead to the temporary accumulation of large lakes of water upstream, so that when the dam breaks a wave of water floods the reaches of river down stream.

d. The collapse of man-made structures across a river can release large volumes of water to produce a river flood. In the 19th century there were several cases in which large earth dams built across rivers failed and released large floods of water that caused great damage immediately downstream.

Because of the potential impact of floods and the danger that they can bring, it has been necessary to develop techniques of flood frequency analysis and flood forecasting. By using data from the period of hydrologic records to establish a relationship between the large discharges and their return period or probability of occurrence, it has been possible to develop techniques of flood frequency analysis which can be used to estimate discharges of specific recurrence intervals. The occurrence of floods can be modified by human activity. Particularly in areas where the drainage basin characteristics of vegetation and land use have been changed, then runoff from the drainage basin can be accelerated so that floods become larger or more frequent. This is particularly the case downstream from urban areas, because the impervious area associated with urbanization gives rise to much greater runoff ad so to increased flooding.

Control measures

The mitigation of river floods can be achieved in four main ways. First, it is possible to bear the loss and do nothing about the damage that results, and this is appropriate in many area of the world where there are few settlements. Secondly, it is possible to modify the losses caused by damage and this can be achieved by using flood insurance, by structural measures (e.g. building houses on stilts), or by zoning land uses so that the least vulnerable are close to the river and the most vulnerable are furthest away. A third approach is to modify the river flood, and this can be done by constructing dams across the river so that if the reservoirs are kept at a level much lower than that of the dam wall, then when a flood comes down the river it simply fills up the reservoir and the water is then released slowly rather than affecting the river downstream of the dam. A debate in the USA in the 1950s centered on the advantages of big dams or little dams, because it proved possible in some drainage basins to construct many small dams in the headwaters of the basin, whereas in other environments it was more effective to construct one large dam along the major river. The fourth method is to use engineering works to contain the flood. In effect, this means modifying the river channel so that it can hold the flood water that comes down.

EARTHQUAKES

Earthquakes are the main seismic hazard. Earthquake caused greater hazards than any other natural phenomena. This is evident from the death of more than 100 million people. Since the beginning of the recorded history of man. In India, more than 60,000 lives were lost during that period. In the year 1897 the Meghalaya earthquake – one of the most disastrous natural phenomena laid waste nearly 380,000 km2 area & killed more than 1600 people in less than a minute. The annual roll of human life all over the world is over 15,000. An earthquake is a sudden and temporary vibration set up on the earth's surface, ranging from a faint warmer to a wild motion, due to the sudden release of energy stored in the rocks-beneath the earths surface. Earthquake is a form of energy wave motion which originates in a limited region and then spread out in all directions from the service of disturbance. Earthquakes usually last for a few seconds to a minute. Sometimes, the vibrations are so feeble that we can not feel them, whereas the violent earthquakes result in huge material loss and the loss of human lives.

The point within the earth, where earthquake waves originate is called the focus and from the focus the vibrations spreads in all directions. They reach the surface first at the point immediately above the focus and this point is called the epicentre. It is at the epicentre where the shock of the earthquake is first experienced. Earthquake occur beneath the surface where rock masses subjected to pressure or stress get strained, leading eventually to breakage along weak zones or faults. Not only is there a sudden stress drop but the strained rocks rebound along the fault so that the stress are suddenly fully or partially released. The breaking of rocks is accompanied by their displacement and the rupture spread along the fault surface at a rate of 2-3 km/s in all directions in a series of uneven movements. This uneven spreading of displacement leads to bursts of high frequency wave which travel in all directions producing seismic shocks. Earthquake emerges at various depth, which may be anywhere in the crust or as far 700 Km down into the mantle.

Causes of Earthquakes

Earthquakes originate due to various reasons which fall into two major categories vis non-tectonic & tectonic

(1) Non-tectonic causes – The non-tectonic causes of earthquake include those associated with the geological agents operating upon the surface of the earth like - Volcanic eruptions as well as with the collapse of subterranean cavities because of the removal of support from below, by action of underground water or sudden subsidence of ground surface.

(2) Tectonic causes – About 95% of all earthquakes are due to sudden earth movements along existing or new faults. The association of earthquakes with fault-lines is an established fact. Earthquake caused by faulting or folding in the crust are known as Tectonic earthquake. The term tectonic refers to the structural changes of the crust due to deformation or displacement. Such earthquakes generally result from sudden yielding to strain produced on rocks by accumulating stresses.

Effects of Earthquakes

Earthquake is a natural calamity such a type that it never gives opportunity and scope to people to save their lives and escape. There are earthquakes almost every year causing large scale damage and devastation. Earthquake take an average yearly toll of 14000 lives and cause damage to extensive property. All the earthquakes are not of equal strength and that the extent of damage depends on the degree of acceleration with which the ground rocks during the earthquake. Damage due to earthquake varies with the strength of the earthquake, local bedrock, type of building construction and life support system. Some of the important effects of the earthquake can be summarized as follows –

(i) Due to vibration of ground, building, bridges, dams, poles and posts and fences etc. may be slightly or heavily damaged and people are hit by falling debris from buildings. Railways are blocked and twisted.

(ii) Earthquakes also cause changes in the geological structure of an area. There may be both vertical as well as horizontal displacement of rocks causing development of slopes or scoops and sometimes fissures and open rocks etc. It may also destroy the road, communication and tear apart the water pipes and gas pipes etc.

(iii) Landslides and subsidence of land also take place during an earthquake. Sudden subsidence of the land near sea or lake cause flooding. In the peruvian earthquake of 1970, million tons of ice, snow, rock and boulders moving at a tremendous speed (estimated at 480 Km/hr) buried the town of Yungay and its inhabitants.

(iv) Ground water and its movement get disturbed by earthquakes, besides, the causes of streams and rivers change new springs develop and in certain favourable conditions sand dykes may also develop.

(v) Fire is a usual problem associated with earthquakes, due to broken gas and water mains and fallen electrical wires.

(vi) Tsunamis an important secondary effect of a major earthquake is the seismic sea wave, or tsunami as is known to the Japanese. An earthquake below the sea floor generates seismic sea-waves which often have catastrophic consequences. They usually devastate the costal regions. Its long duration & great height causes great damage to the entire coast and many deaths by drowning in low-lying coastal areas. It is through that coastal flooding which occurred in Japan in 1703, with an estimated life loss of 1,00,000 persons, may have been caused by seismic sea waves.

Control of Earthquake

Earthquake proof construction – Since the damage caused by an earthquake depends on the rate of vibration of the quake, the construction should be made accordingly. It has been seen that if the structure is made quite firm, the shocks may not be able to cause much damage to it and the structure can withstand the vibrations. In such buildings the roofs are made as light as possible. However, in regions where earthquake are frequent, the structure should be made of lighter materials like wood.

Some Important Earthquake in India

1- Kutch Earthquake of June, 16th, 1819

2- Bengal and Kashmir Earthquake of 1885

3- Assam Earthquake of 1897, 1935, 1950, 1988 (6th Aug. 1988)

4- Kangra Eathquake, 1905, 1975, 1987

5- Bihar Earthquake, 15th Jan 1934, 21st Aug. 1988

6- Koyna Earthquake, 11th December, 1967.

7- Uttarkashi, October 1991.

8- Latur, Nov 1993.

9- Chamoli May 1999.

10- Bhuj, Jan 2000.

11- Kashmir, 2005

CYCLONE

Tropical cyclones, at times, bring nature's worst disaster in the tropics. Slightly away from the equator within the belt of tropics (between 30° N and 30°S) the tropical cyclones are formed everywhere over the ocean where the ocean water is warm. However, their frequency, intensity and coastal impact vary from place to place. The frequency of tropical cyclone is the least in the North Indian Ocean (the Bay of Bengal and the Arabian Sea) and they are mostly moderate in intensity, but they are the deadliest when cross coast bordering the areas of North Bay of Bengal (coastal areas of North Orisa, West Bengal and Bangladesh). Tropical cyclones are the intense low pressure areas in the atmosphere around which fierce with blows. Horizontally, it extends from 500 to 1,000 km and vertically from surface to about 12 to 14 km. On an average, about five-six tropical cyclone from in the Bay of Bengal and the Arabian sea every year out of which two to three may be severe. More cyclones occur in the Bay of Bengal than in the Arabian sea and the ratio of their frequencies is about 4:1 May, June, October and November are the stormiest months of the year. Compared to the pre-monsoon season, particularly the months of October and November are known for severe storms.

Severe tropical cyclones are responsible for large casualties and considerable damage to property and agricultural crop. The destruction is confined to the coastal districts and the maximum destruction being within 100 km from the center of the cyclones and on the right of the storm track. Principal dangers from a cyclone are: (i) Gales and strong winds, (ii) Torrential rain and (iii) High tidal ways (also known as 'Storm surges') Most casualties are caused by coastal inundation by tidal waves and storm surges. Maximum penetration of severe storm surges varies from 10 to 20 km in inland from the coast. Heavy rainfall and floods come next in order of devastation. They are often responsible for much loss of life and damage to property. Death and destruction purely due to winds are relatively less. The collapse of buildings, falling trees, flying debris, electrocution, rain and aircraft accidents and disease from contaminated food and water in the post-cyclone period also contribute to loss of life and destruction of property.

The cyclone warning service of the Indian Meteorological Department is now over a hundred years old and is one of the most important functions of the Department. At present the cyclone warnings are provided through the Area Cyclone Warning Centres (ACWCs) located at Calcutta, Madras and Bombay and Cyclone Warning Centres (CWCs) at Bhubaneshwar, Visakhapatnam and Ahmedabad. The zone of responsibility of each office is clearly demarcated. Cyclone warning bulletins for All India Radio/ Doordashan and cyclone advisories for the north Indian Ocean to Bangladesh, Burma, India, Maldives, Pakistan, Sri Lanka and Thailand are being issued from Meteorological Office at New Delhi. This office in New Delhi also issues tropical cyclone advisories for the tropical cyclones in the south-west Indian Ocean to Mauritius. The important components of cyclone warnings are the forecast of future path, intensity and the associated destructive weather systems. For the preparation of forecast of path and storm surges the modern methods, which utilize computers are used in addition to conventional methods. For the intensity forecast satellite techniques are used.

LANDSLIDES

Landslide and mass movement are recurring phenomena in Himalayan region. The consequences in recent times have become more severe in terms of casualties and extensive damage to the roads, buildings, forest, plantation, and agriculture fields. In recent years the intensive construction activity and destabilizing forces of nature have combined to generate a class of problems, huge and complex, never encountered before. Implementation of number of hydro-electric schemes, large scale construction of dames, roads, tunnels, buildings, towers, ropeways, tanks and other public utility works as well as indiscriminate mining and quarrying have brought most of instability problems such as never witnessed before.

It is well recognized that landslides, are of appreciable significance in the mountain areas of Uttarkhand. Knowledge about landslides identification and prevention was largely lacking in early 1970's some public works were built on an old landslides deposits, such activities reactivated some landslides comes heavy damage of life and property. Landslides are a geomorphological process influenced by morphometric features such as slopes, valleys, escarpments etc. These are formed through endogenous as well as exogenous forces. Landslides occur in different ways and on various scales. Anthropogenic interference's with the environment is another factor responsible for landslides. Sometime development schemes such as construction of dams or reservoirs, housing scheme, roads, agricultural practices in steep slope, etc. implemented without proper environmental impact assessment can also precipitate landslide.

Landslides, which occur as a consequence of changes in landforms, the causes of such changes are preventive as well as remedial measures that could be adopted, all come within the preview of environmental geomorphology. The formation as well as the destruction of land forms in every natural process. But a combination of forces in at a work here. A landslide are associated with hill-slopes, rain is another contributory factor. But neither all hill slopes nor all rain-affected areas are subject to landslides. In many places in Uttarkhand major landslides occur due to blasting for road cuttings. Landslides can be grouped into three categories.

- Those which are unpredictable, which cause heavy loss of life and extensive damage example –Karmi landslide in 1983 and very recent landslide of Berinag in 1996.
- Those whose threat is known.
- The Third category are landslides which are likely to occur associated with a proposed development project such as a dam or reservoir housing or a new road.

Control Measures

A significant reduction of landslides hazards can be achieved by preventing or minimizing the exposure of populations and facilities to landslides and by physically controlling landslides that occur. The problem of management of landslides through effective control measures would require enormous financial resources which will never be available. Hence the wisdom lies in full exploitation of human resources and in making best use of local skills and materials.

The following measures may be under taken to prevent further landslide.

Drainage Measure - Surface drainage

- Sub surface drainage

Erosion control Terracing - Bamboo check dams

- Jute and coir netting
- Rockfall control measure
- Grass plantation
- Vegetated dry masonry wall
- Retaining wall
- Afforestation.

The control works that are actually carried out in the landslides areas are primarily for the purpose of saving life, secondly for the preservation of public structures and buildings and thirdly to prevent the disruption of road traffic and to prevent flooding in the event of a landslide damming a river.

DISASTER MANAGEMENT

The problems of disaster management particularly in developing countries, such as India, are unique due to the seemingly competing needs of basic necessities for people and economic progress. It is believed that the unplanned march of development activities at the cost of natural environment has made most parts of the country susceptible to a fatal mix of forces of nature and human error. The alarming increase in population, depletion of resources, rising economic disparities and unprecedented scientific and technological development are considered to be the factors that are mainly responsible for wide spread vulnerability of the society and its economic, political and organizational structure to the onslaught of disasters.

Disaster management is an applied science that seeks by the systematic observations and analysis of disasters to improve measures related to the prevention, mitigation, preparedness, emergency response and recovery. Thus whenever we talk of disasters, we invariably imply separate phases or preventive measures for ensuring zero disaster state of preparedness and follow up actions in the event of occurrence of disaster. Thus a well-formulated disaster management plan comprises the following different phases:

1. **Disaster Prevention:** This includes all those tasks that can be undertaken to prevent a natural hazard from turning into a disaster. Though not much can be done to prevent the natural hazards like cyclones, floods and avalanches but efforts can be made to prevent their calamitous fall out. The preventive measures may be taken under the category of national development and also under the specific disaster management program.

2. **Disaster Mitigation**: Disaster mitigation means taking action to reduce the effects of hazards before it occur. The term mitigation applies to wide range of activities and protection measures that might be instigated, from the physical, like constructing stronger buildings, to the procedural, like standard techniques for incorporating hazard assessment in land use planning. It is a major component of disaster management plan. Watershed management, channel improvement, alternative cropping pattern, live stock management and soil conservation techniques are also important components of disaster mitigation. These strategies would vary from disaster to disaster. The non structural measures include adequate legislation, rules and by laws to guard against the disasters, insurance and microcredit schemes, education and training programs, activation of self help groups, suitable warning system and institution building. A composite multisectrol and long term health plan is also an essential part of disaster mitigation.

3. **Disaster Preparedness**: Planning is one of the most efficient tools available to deal with disasters. The preparedness measures to withstand disasters must fall within sub plans on specific disasters at the central, state, district and block levels. A disaster preparedness plan should be clear, realistic, flexible, viable and easy to use. A preparedness plan must cover every stage of

disaster management cycle. There is also a need for separate contingency plans, relief plans and recovery plans for droughts, floods, cyclones and earthquakes.

Identification of vulnerable areas and section should be the first step of disaster preparedness efforts. Adequate vulnerability capacity assessment (VCA) will improve the efficacy of the disaster preparedness plan. During a disaster, a VCA will help in rescue and relief work and after a disaster, it will be able to determine the need, type and extent and length of assistance. The VCA is an important tool that can be used for ascertaining physical, social and economic vulnerability designing disaster management program and monitoring and evaluating the disaster impact. Collection and prediction can help in managing disasters more effectively. With the insight provided by socio-economic vulnerability analysis, it is possible to estimate the direct and indirect loses and mitigate their impacts.

Since communication is the first casualty, a satellite based dependable and unique communication system known as Disaster Warning System has been developed for coastal areas. Now low-pressure area can be tracked 72 hours in advance for effective management of disasters. We need an adequate, reliable and timely warning system that is possible only on the basis of reliable forecasts. Accurate forecasts can be made only for those disasters that are predictable. There is, therefore, a significant link between predictability, forecasting components of disaster preparedness planning and need to be supported by a good public information policy.

Strengthening the organizational set up to keep a watch on disaster management programs is another crucial area that requires due attention. Some of the notable agencies and departments, which could play important role in identifying the major disaster threats, and in suggesting as well as implementing preventive measures are:

1. Government Ministries and Department (National Crises Management Committee, Crises Management Group, Ministries of Agriculture, Water Resource, Health, Railways, Environment, Atomic Energy, Information and Broadcasting, etc.).

2. Academic and Training Institutes (India Institute to Public Administration, Indira Gandhi National Open University, Yashwant Rao Chavan Academy of Development Administration, University of Rookie, etc.).

3. Research Organization (Forest Research Institute, Defense Research and Development Organization, Center for Science and Environment, etc.).

4. Scientific and Technical Organizations (Council for Scientific and industrial Research, India Space Research Organization).

5. Non-Government Organizations (Disaster Mitigation Institute, Ahmedabad, Barh Mukti Abhiyan, Bihar, Kalpvriksha, Pune).

6. International Agencies (World Bank, International Red Cross and Red Cross Societies, UNDP, UNESCO, UNICEF and World Health Organization).

4. **Disaster Response:** The effectiveness of Disaster Preparedness Plans depends largely on systematic damage assessment process. Damaged assessment is a prerequisite for all disaster management activities and precedes the disaster response work. This type of assessment is needed for short-term emergency relief measures as well as long-term restoration and recovery work. Important steps that area required to be taken up by the concerned agencies to provide disaster relief to the affected communities include search, rescue and evacuation tasks, provision of shelter and power supplies, food security (food subsidies, distribution and storage), clearance of debris, establishment of communication net work, appropriate health and

sanitation facilities, public information, security and welfare mechanism. Most importantly, warehousing and stockpiling of essential items for distribution of relief material are critical to any disaster management exercise.

Livestock management is the most neglected area of disaster response programs. An effective livestock improvement program should, therefore, include establishment of animals shelters in disaster affected areas, provision of water, medicines and injections, training to livestock relief workers and veterinary doctors, removal of dead animals and carcasses and establishment of all long-term disaster management tasks. The psyche of the provider as well as the sufferer needs to be taken into consideration at this stage itself as it paves the way for operative recovery programs.

5. **Disaster Recovery**: Recovery stage of disaster management cycle covers rehabilitation and re-construction activities. Disaster recovery is a slow process, it begins in the aftermath of disaster and goes on well beyond the culmination of disaster response stage. Recovery activities usually relate to economic, social and infrastructural rehabilitation. Economic rehabilitation lays emphasis on introduction of conducive agricultural development programs, employment generation schemes and promotion of activities such as agro forestry, soil conservation, water resource management, small-scale industries and labor intensive schemes.

Social rehabilitation is the most crucial task as it takes care of the psychological aspects of 'trauma' that is the foremost but inconspicuous outcome of disasters. It includes activities related to re-activation of anganwadis, community centers, female/children homes, and old age homes. The objective should be to pay extra attention on single parent families, women children, elderly and handicapped. Resurrection of educational and training activities should take place immediately after the major part of the relief work is over. The vulnerable should be rehabilitated in familiar environs.

The stress mitigation counseling programs can be very helpful for this purpose.

Infrastructural rehabilitation is dependent on creation of additional facilities within the existing health institutions, educational and researches centers. Construction of cyclone and earthquake resistant buildings should be promoted. Strengthening, retrofitting and reconstructing of houses should also form a part of the rehabilitation endeavors. Efforts must be made to build up alternative power and communication network in the disaster affected regions as early as possible.

The areas where in spite to best mitigation efforts, disasters do strike may also be viewed as areas where opportunities always exist to reconstruct and rebuild the entire socio-economic infrastructure.

The sequence in disaster management cycle is incomplete if the issue of disaster and development is not addressed properly. It is contended that development that is unplanned and unmindful, precipitates disaster conditions. While, the other side of the picture, speaks of the opportunities that are thrown open for development programs in the aftermath of disasters. It is indeed a chicken and egg problem. Development does aggravate disaster conditions if not create them altogether and at the same time disaster recovery activities have often led to promotion of development activities in the affected area. It must be remembered that destruction caused by disaster can only be directed towards development if the agencies (Government, non-government, Military, Para-military) active in this field work in unison. Good leadership can make the task simpler and easier. Coordination of the activities of all concerned bodies is needed to avoid wastage of resources and duplication of efforts. The success of disaster management programs depends not just on the effective planning execution of disaster relief activities at different stages of disaster management cycle but also on the involvement of an alert and informed community in the disaster related activities.

The most recent earthquake of China

The 2008 Sichuan earthquake or the Great Sichuan Earthquake, which measured at 8.0 and 8.3 according to China

Seismological Bureau (CSB), and 7.9 according to USGS, occurred at 14:28:01.42 CST (06:28:01.42 UTC) on 12 May 2008 in Sichuan province of China. It was also known as the Wenchuan earthquake after the earthquake's epicenter in the Town of Yingxiu , Wenchuan County, Ngawa Tibetan and Qiang Autonomous Prefecture, Sichuan province. The epicenter was 80 kilometres (50 mi) west-northwest of Chengdu, the capital of Sichuan, with a depth of 19 kilometres. The earthquake was felt as far away as Beijing (1,500 km away) and Shanghai (1,700 km away), where office buildings swayed with the tremor. The earthquake was also felt in nearby countries. Official figures state that 69,181 are confirmed dead, including 68,636 in Sichuan province, and 3,74,171 injured, with 18,522 listed as missing. The earthquake left about 4.8 million people homeless, though the number could be as high as 11 million. It is the deadliest and strongest earthquake to hit China since the 1976 Tangshan earthquake, which killed at least 240,000 people. Approximately 15 million people lived in the affected area.

On May 25, a major aftershock of 6.0 M hit northeast of the original earthquake's epicenter, in Qingchuan County, causing eight deaths, 1000 injuries, and destroying thousands of buildings. On May 27, two more major aftershocks, a 5.2 M in Qingchuan County and a 5.7 in Ningqiang in Shaanxi Province, led to the collapse of more than 420,000 homes and injured 63 people.

CHAPTER-XIII

Social Issues and The Environment

I. FROM UNSUSTAINABLE TO SUSTAINABLE DEVELOPMENT

When in the decade between 1960-70 environmental isues came to the fore front, a common question came in to the mind of many that how long will it be before we reap consequences of our environmental disregard? This question is very pertinent even today after almost 40 years. As the world has already started experiencing a hotter climate, holes in the protective ozone shield, various toxic chemicals in the soil and ground water, food contaminated with pesticides residues, diminishing forests, increasing natural and man made disasters , famines, and extinctions of species from the surface of earth. And perhaps we do not have another 40 years to sustain life on this planet earth. The various developmental activities of human beings such as industrial revolution, agricultural revolution and cultural revolution all have casused environmental degradation. Though these developmental activities are essential for survival of man, it is necessary to take suitable environmental protection measures to contain the further environmental degredation . Over the next 40 years we must either create a sustainable society or bear the consequences such as a deteriorating quality of life as air and water, natural environments, and number of wild species are increasingly destroyed by our polluting wastes and lust for exploitation.Also more and more people will become environmental refugees, depending for help on others because the depleted soil and water resources no longer will support a productive livelihood. Development at the rampant destruction of environment can only be a short term one. In the long term it becomes an anti development and can go on only at the cost of enormous human sufferings, increased poverty and oppression. It has, therefore, become necessary to understand the processes of environment for a sustainable development.

Sustainable development means the development that meets the needs of the people today without compromising the ability of future generation to meet their own needs.

According to World Commission on Environment and Development (WCED), this is development that meets the needs of the present without compromising the ability of future generations to meet their needs. Sustainable development implies economic growth together with the protection of environmental quality, each reinforcing the other. The essence of this form of development is a stable relationship between human activities and the natural world, which does not diminish the prospects for future generations to enjoy a quality of life at least as good as our own. Many observers believe that participatory democracy, undominated by vested interests, is a pre-requisite for achieving sustainable development.

One of the fundamental idea of sustainability suggests that the under lying conditions necessary for success in biological enterprises will be maintained. This involves the maintaining the sources of seeds and cultivars, maintaining fertility and habitat quality and keeping nourishing geophysical systems in good repair.

Thus development should be a process for the utilization of different resources weather renewable or non-renewable, for the purpose of providing better services to human beings of the present day and to the generations yet to come. This goal can achieved through the sound utilization of environmental resources.

Modern development in its short life, has shown great tendency towards global resources depletion. It has not only be non responsive to employment needs of the labour force but has led to unemployment specially in the developed countries. Despite the emphasis on development, the gap between rich and poor countries has widened. The sound austainable development of the environment and various resources with due consideration for the physical, environmental, social, cultural, ethical and spritual needs of human beings is the alternative which must be followed in the context of long established human-environment relationships.

Development should not alienate humanity from the environment but bring it to terms with it. Thus in order to achieve lasting solutions, we must refocus our attention and priorities on the principles underlying sustainability. These are:

(i) Stabilizing population

(ii) Attaining a sustainable agriuculture, one which does not deplete soil and water resources or contaminate the eart and food supplies with pesticides

(iii) Recycling Wastes

(iv) Developing benign solar energy alternatives

(v) Adopting more energy-and resource conservative life style.

Commission on Sustainable development

In 1992, more than 100 heads of States met in Rio de Janerio, Brazil for the United Nations Conference on Environment and Development (UNCED). The earth Summit, as the UNCED was also known, addressed the urgent problems of environmental protection and socioeconomic development. The leaders signed the Frame work Convention on Climate Change and the Convention on Biological Diversity. The United Nations Commission on Sustainable Development was created in December 1992 to ensure active follow up of UNCED. The CSD is a functional commission of the UN Economic and Social Council(ECOSOC), with 53 members. A five year review of earth summit progress took place in 1997, followed in 2002 by a ten year review of World Summit on Sustainable Development. At the eleventh session of CSD held in New York from 28th April to 9th May 2003, decisions were made on the commission's future programme and organization of work.

The Commission on Sustainable Development is an inter governmental body with 53members elected for a period of three years, meets annually for a period of two to three weeks. The members are elected from amongst the member states of the United Nations specialized agencies. 13 members are elected from Africa, eleven from Asia, 10 from Latin America and the Caribbean, 6 from Eastern Europe, and 13 from Western Europe and other. One third members are elected annually and outgoing members are eligible for re-election.

Sustainable Development in Developing Countries

The Republic of Indonesia is the fourth largest country in world in terms of population and consists more than 13,000 islands. Bali, is one of the 27 provinces of Indonesia but only 5,600 Km^2 in area, is one of the best known places in the country. Bali faces many challenges as it tries to encourage vigorous economic development while respecting the integrity of the culture of its people and physical environment. Since the early 1970s, after growth of international tourism, Bali faces negative impacts of tourism like scarce water supplies unwillingness of Balinese to work in agriculture sector. So, to cope with such a situation Government used five year plans to guide development which incorporated various criteria to maintain ecological integrity, efficiency, equity, cultural integrity, community participation and development as realization of potential means to enhance the capabilities at all levels from village to the district, province and to the national needs and improve quality of life.

II. URBAN PROBLEMS RELATED TO ENERGY

All the environmental problems we have been facing, ranging from degradation of agricultural soil and water resources through the various forms of pollution to depletion of traditional sources of enerfy have their origin with the people and their life style. Farms and natural areas surrounding cities are continually giving way to new housing developments, shopping centers, industrial parks, parking lots, and other facilities interconnected with new highways carrying streams of traffic. This rampant growth is acceptable and promoted as an indication of a strong economy. However, it is largely unplanned and uncontrolled and lies at the root of our environmental problems. Despite their virtues, walking and public transit did not make cities desirable place to live. Specially in industrial cities, poor housing, inadequate sewage systems, inadequate refuse collection, pollution from industries, and generally congested, noisy comditions combined to make cities trying places to live. The development proceeds according to the whims of individual developers, governed only by where they could acquire land as opposed to any over all planning.

The urban sprawl life style translates in to an increasing demand for energy resources, specially crude oil and electricity needed to fuel motor vehicles as the number of cars and motor vehicles and commuting miles have increased and for industries and houses . Thus the problems and consequences of oil shortages that we fae are a direct consequence of urban citizen. Like wise individual sub-urban homes require 1.5 to 2 times more energy for heating and coolong than comparable attached city dwellings. Therefore a large part of the growth in demand for fuel oil, natural gas and electrical power, which is largely produced from coal or nuclear power plants, is a result of the shipt to more spacious sub-urban living.

Energy has played a very crucial role in fostering growth of cities. The physical spread of settlements was previously based on somatic energy but with the discovery of ways to generate cheap and abundant energy, settlements grew in large areas. For example, even major cities in pre-industrial period spread over only 3 to 5 sq km and were easily traversed on foot like London (320 ha), Paris (370 ha) and Delhi (500ha) etc. However, Industrial revolution encouraged massive migration to industrial centers, which also happened to be major cities and they grew rapidly in population as well as in area. In India, the phenomenon of rapid population growth in urban centers has been witnessed from 1941 on wards. Most cities have quadrupled their populations in the last 50 years. Cities have spread horizontally as well as vertically, many extending over 40-50 sq km. They required mechanized transport, consuming vast amount of commercial fuels. Expansion also reduced the ability of the urban fabric to dissipate heat to surrounding countryside. The cities thus develop as a reservoir of heat which necessitated consumption of commercial energy to achieve the same degree of comfort. Industrial processes require vast amounts of commercial energy for production which add to the wealth of the cities as well as its miseries by polluting environment.

In a city, energy is consumed by people for various activities and the type of consumption may be recurring, non-recurring or both. Recurring energy is consumed in urban areas in domestic sector – for cooking, heating cooling, recreational purposes. Transport sector includes distribution & transportation of goods and social visits. Water supply, sewage and drainage, incineration all these require energy. Non recurring consumption of energy takes place through construction of buildings and infrastructure networks. Extraction of raw materials also requires energy. Material s required for modern construction such as bricks, cement, steel and metals like copper, brass or aluminum and PVC require high consumption of commercial energy for their manufacturing process. According to a study by Central Building Research Institute (CBRI) on energy requirements of different building materials, following amount of energy is required in their manufacture:

Bricks	-	1.02×10^3 Kcal/kg
Cement	-	1.93×10^3 Kcal/kg
Mild Steel	-	8.3×10^3 Kcal/kg
PVC	-	27.75×10^3 Kcal/kg
Aluminum	-	34.3×10^3 Kcal/kg

The pattern and extent of energy consumption of cities varies between city to city.

III. WATER CONSERVATION, RAIN WATER HARVESTING AND WATERSHED MANAGEMENT

Water is considered an inexhaustible natural resource and conservation means rational use of a resource. Water has assumed the status of a strategic resource and one upon which economic development is predicted. Drought and general water shortage has led to a renewed interest in water resources conservation. In the developed countries continuous programs for renewal of the distributional infrastructure and well designed pricing policies have usually reduced the demand of water to a manageable level. Good conservation methods of irrigation, such as drip irrigation can reduce problems such as evaporation and infiltration losses, salinization and water logging. Mulching and spraying of chemical films on fields can also reduce evaporation losses.

WATER CYCLE

The water cycle, also known as Hydrological cycle describes the circulation of water from earth to atmosphere, and vive-versa. The cycle is composed of two phases, viz., phase (i) consists of various processes by which water is available to the atmosphere from the earth; and phase (ii) consists of various mechanisms associated with the condensation of water vapor, and its availability to the earth in the form of precipitation.

Phase I. Evaporation, Sweating and Transpiration

These are the principal mechanisms by which water is available to the atmosphere, in the form of water vapor. The water from the seas, river, lake, pond, and other water bodies evaporates under the influence of solar radiation into the atmosphere. Animals lose water through sweating and by evaporation through the integuments, during respiration, via urine and in the feces. Moisture evaporates directly back to the atmosphere from the soil surface. Soil water is also absorbed through plants, and is returned to the atmosphere via transpiration. Thus, various features on the earth constantly keep on evaporating and transpiring water into the atmosphere in invisible water vapor form. This invisible water vapor (which is lost from the water bodies through evaporation and transpiration) is carried over the continents by moving air masses. The water vapor moves in accordance with the wind direction.

Phase II Precipitation

Precipitation is the sources of all fresh waters on the earth surface. It is a result of gravitational pull on the condensed water vapor of the atmosphere. Lower temperature is required for condensation of water vapor into water droplets. The temperature, at which condensation occurs is known as the dew point. This point varies with the moisture content of the atmosphere. Some important forms of precipitation that frequently occur in nature include:

(i) Drizzle : Uniform precipitation of minute drops.

(ii) Rain : Precipitation of liquid water in drops larger than drizzle.

(iii) Snow : A form of precipitation in the solid stage. Snow fall is a characteristic feature of high mountains and of polar regions.

(iv) Hail : Precipitation in the form of balls or lumps of ice.

(v) Sleet : A type of precipitation in the form of grain or pellets of ice.

(vi) Frost or Dew : Dew is the form of precipitation, which is formed as a result of condensation of water vapor on the surface of the object. When surface temperature of the object is below freezing point the water vapor condenses as a frost. Dew may be important locally in arid regions.

Of all these, over most of the earth's surface, the most important form of precipitation is rainfall. For the production of a significant precipitation, four mechanisms are necessary. These are :

1. Uplifting and Cooling of air

A vertical lifting mechanism produces cooling of the moist air in the colder belts of atmosphere. Spatial and temporal variations in vertical motion of the air, largely determine the variation of precipitation in space and time. These motions give rise to significant rainfall. In fact, the vertical motions of the air are associated with weather systems such as thunderstorms (lighting and thunder with heavy rain), cyclones (a system of winds blowing spirally inwards, towards the center of low pressure, and rise up vertically). Uplifting of air masses is also caused by physical barriers like mountain ranges. There are three main types of vertical lifting of air masses in the atmosphere, which are responsible for the significant rainfall. These are: (i) Convective lifting (resulting in precipitation over a local and is only of short duration), (ii) Banded lifting (resulting in precipitation over a zone), and (iii) General lifting (associated with large scale weather systems, and resulting precipitation extends over the large area.). In India, significant precipitation in the form of rainfall is of a special pattern of moist air motion known as the monsoon (monsoon is a periodical wind of the Indian ocean, south-west from April to October and north-east the rest of the year).

2. Condensation of water vapor and formation of precipitation

Condensation of water vapor in the atmosphere takes place on small particles having an affinity for water, known as condensation nuclei. These small particles are in fat, aerosols suspended in the atmosphere. The size of aerosols varies from molecular to 100 mm. Aerosols are usually characterized in part by the terms like : (I) Background or natural aerosols, (ii) Tropospheric aerosols, (iii) Stratospheric aerosol, (iv) Maritime aerosols, and (v) Continental aerosols. An undisturbed natural environment always consists of natural aerosol. However, the natural environment may include particles from anthropogenic sources (in the past 50 years, there has been about 100 per cent increase in its level in the Northern Hemisphere; in 1972 it amounted to about 300-500 particles/cm3, it varies considerably near its sources, but beyond 5 kilometers or so above the ground, it is almost steady). The other terms refer to aerosols (Tropospheric, Stratospheric, Maritime, and Continental) are specific to certain levels or locations. The aerosols are formed by one or more of the following process:

1. Wind effects on the surf and so on.
2. Interactions during scavenging processing.
3. Photo chemical reaction between trace gases.
4. Absorption of gases in particles or droplets with subsequent chemical reactions.

The aerosol particles may be inorganic, e.g., NaCl $(NH_4)_2 SO_4$, volcanic ash, clay material, or they may be organic, e. g., shoot bacteria, or oil globules. The natural aerosol is characterized by sulfates with ammonia. It has been found that above 50 per cent relative humidity in the atmosphere, most of the aerosol particles contain some water on their surface due to their hygroscopic nature. At 80-90 per cent relative humidity, the wetted particles began to affect visibility, and haze or smog layers are formed. At 100 per cent relative humidity (or slightly above), fogs or clouds are formed. The condensation nuclei are the special particles that serve as the condensation site on which each fog or cloud droplet forms at the relatively low super saturations (usually $<$ 1 per cent present in clouds or even lower supersaturating < 0.1 per cent in fog). There are always sufficient nuclei present in the lower atmosphere for condensation to occur if the air is cooled to saturation, and in the smoky industrial areas before the saturation point is reached. If fact, the saturation of the air in the presence of sufficient condensation nuclei near the earth surface (land or water) causes the formation of fog. Fog is defined as 'a cloud that envelops the observer and restricts his horizontal visibility to 1,000 meters or less'. Fogs are classified according to cause of their formation into the following types:

(i) Advection fog

These are coastal and open water phenomena. Warm moist air moving over a water surface that becomes colder down wind, provides the conditions favorable for its formation.

(ii) Radiation fog

Valleys are particularly characterized by radiation fog. Air that is cooled at higher elevations drains into the valleys where it accumulates, resulting in dense fog. The radiation cooling (stagnant moist air near the ground becomes progressively cooler during a cloudless night because of an excess of out going radiation) further lowers the temperature of the air.

(iii) Upslope fog

It is the air mass fog which results when stable air is adiabatically cooled to its saturation point, and is uplifted.

3. Growth of cloud droplets to sizes capable of falling to the ground

All clouds do not produce rain. Most clouds evaporate without even producing precipitation at the earth surface. For an active precipitation ground is essential (clouds are suspensions of minute liquid water drops being of the order of 1-100 mm in diameter, while the typical raindrop has diameter of 1 mm). Therefore, it is essential to explain the growth of cloud droplets to such a size that they can fall against the rising air, and produce rain. There are two theories to explain the processes :

(i) Bergeron theory or Ice-crystal process

This theory explains that the clouds on the middle latitudes consist of many ice-crystals in a close proximity to water droplets. The water vapor moves from droplet to ice crystals due to more saturation vapor pressure over water than that over ice(maximum difference in vapor pressure occurs in the range of –10 and 20°C). Now the ice-crystals are known as snowflakes. The behavior of snowflakes for precipitation depends on the temperature of lower atmosphere. If air in the lower atmosphere is at or below 0°C, the precipitation occurs in the form of snow, on the other hand, when air is warmer in form of rain. Such behavior of snowflakes can easily be marked on the mountains with increasing altitudes.

(ii) Coalescence Theory

This theory explains that in the tropics the clouds do not contain ice-crystals being completely above 0°C temperature. The clouds contain two types of original water droplets : (I) larger, and (II) smaller. The larger droplets are originated from the salt particles which, intern, originate by evaporation of spray from sea surface. Presence of larger droplets in the cloud causes slower rate of rise, and eventually leads to collision and coalescence with smaller droplets. The growing (due to collision with smaller droplets) and uplifting cloud ensures that the drop will not evaporate at the top of the cloud, but will fall back through the cloud (which is growing further by collision), resulting precipitation in the form of rain.

4. Accumulation of water vapor in the storms or behavior of shower clouds

Behavior of shower clouds is the precipitation efficiency (PE), defined as 'the ratio of the mass of precipitation reaching the ground to the total mass of water vapor passing upwards through the cloud base'. The PE is zero in small cumulus clouds. Small storms observe about 10 per cent PE, however, its excess 50 per cent for large thunderstorms. In fact, the PE is influenced by the flow of moist or dry air into the precipitating system. If the inflowing air is dry, the rainfall is small, while if it is very moist, the rainfall is extremely heavy. It indicates that the precipitating system requires a continual flow of water vapor from lower atmosphere to produce effective precipitation. Thus, it is important to note that a strong inflow of very moist air into the storm is essential for the production of continuous intense precipitation.

The above discussion indicates how water from the evaporating and transpiring bodies moves into atmosphere, and how atmospheric water returns back as precipitation. Let us discuss global water balance, and annual flux rates for the world hydrological cycle. It must be remembered that the events of hydrological cycle show great variation both in space and time. Various anthropogenic activities have already produced basic changes in the natural cycle.

GLOBAL WATER BALANCE AND ANNUAL FLUX RATES

Since time to time, attempts have been made by many workers (Hutchinson, 1957; Kalinin and Bykov, 1969; Budyko, 1974) to set-up balance sheet and flux rates for the worlds water cycle. It has been assessed that most of the worlds water is combined chemically with mineral of the primary lithosphere and sedimentary deposits. Indeed, this water is not available easily due to very slow geological weathering processes that break down the rocks. The most of the available water is in the oceans (amounting between 1,32,20,00, 000 and 1,27,00,00,00 cu kms) and the continents (amounting between 3, 78, 37, 000 and 8, 98, 14,200 cu kms). The total amount of water occurring as vapor in the atmosphere (between 123,000 and 14,000 cu kms) is relatively very less compared to the total amount of water on the earth (continents + oceans). At any one time, less than 0.001 per cent of the earth's available water is present in the atmosphere. The atmospheric water which originates precipitation is renewed by evaporation from the open water surfaces and by the transpiration from plants (evaporation + transpiration = evapotranspiration). It has been estimated that between 37,000 and 46,000 km3 of water per year is contributed by oceans to continents via atmosphere. The same amount returns each year from continents to oceans through runoff. It is evident from following figure that over the oceans, the rate of evaporation is greater than of the precipitation. On the other hand, over the continents the rate of precipitation is greater than that of evapotranspiration. This difference between evaporation and precipitation rates over oceans and continents represents the net transport of moisture from ocean to continents, and net discharge as runoff from continents into the oceans. In other words it can be said that the continents receive more water as precipitation than they lose by evaporation. This excess quantity flows over or under the ground surface to reach the sea as runoff. It has been also estimated that on a world wide scale, the moisture coming from the ocean makes up only about 40 per cent of the precipitation falling on the continents. All the remainder (about 60 per cent) moisture over the continents is provided by evapotranspiration from the continental surfaces as plant cover, soil surface, etc. The total annual world-wide precipitation ranges between 5,17,00 and 5, 20,000 km3, while the atmosphere contains water only between 13,000 and 14,000 km3. When total annual precipitation is divided by the quantity of water in the atmosphere at any one time (amounting between 13,000 and 14,000 km3), the so obtained figure (ranges between 37 and 40) estimates the rate of turnover, Here the figure 35-40 (5,20,000/14,000=37, and 5,17,000/13,000=40) indicates that the water from the atmosphere is turned over about 37-40 times a year, i.e., the water vapor in the air is completely replaced every 9-10 days (37-40 times a year means once every 9-10 days). In other words, average residence time of water in the atmospheric pool (reservoir) s 10 days.

The amount of water held in the atmosphere at any given time, in the form of water vapor is very small in comparison to the storage capacities of continents and oceans. Thus, the main hydrological function of the atmosphere is to transport the water vapor. In fact, over a number of years, there is an exact global balance between evapotranspiration and precipitation, i.e., total evapotranspiration (from ocean and continents) equals total precipitation over oceans and continents.

RAIN WATER HARVESTING

As the level of ground water is decreasing on one hand and on the other hand the demand of water for different purposes is increasing, it has become imperative not only to recharge the ground water level but also to cater the needs of the communities by harvesting the water which we receive in the form of rains. Rain water harvesting is the technique through which rain water is captured from the catchments and stored in reservoirs. Harvested rain water can be used for ground water recharge by adopting artificial recharge techniques or to meet the household needs through storage in tanks. It has become very essential to harvest the rain water because of the following reasons:

(a) To meet the ever increasing demand of the water.

(ii) To reduce the runoff which chokes the storm drains.

(iii) To avoid flooding of roads and other areas.

(iv) To augment the ground water storage and control decline of water levels.

(v) To improve the quality of ground water in aquifers.

(vii) To mitigates the effects of draught and achieve drought proofing

(viii) To reduce the soil erosion as the surface runoff is reduced

(ix) To save energy: to lift ground water, one meter rise in water level saves about 0.40 kilo watt hour of electricity.

The amount of water harvested depends on the following factors:

(b) The frequency and intensity of rainfall

(ii) Characteristics of the catchments

(iii) Water demands of water

(iv) How much run off occurs

(v) Rate of infiltration of water through sub-soil and to percolate down to recharge the aquifers.

Design considerations: The three most important components, which need to be evaluated for designing the rain water harvesting structures are-

(i) Hydrology of the area including nature and extent of aquifer, soil cover, topography, depth of water levels and chemical quality of ground water.

(ii) Area contributing for run off i.e. how much area falls under the catchment and land use pattern, whether industrial, residential or green belts and general built up pattern area.

(iii) Hydro-meteorological characters viz. rainfall, duration, general pattern and intensity of rainfall.

The recharged structures should be designed based on-

(a) availability of space

(b) availability of runoff

(c) depth of water table

(d) Lithology of the area

Assessment of runoff (the quantity of water to be collected):

Runoff (m3) = catchment area(m2) X average monsoon rainfall(m) X runoff coefficient

Runoff coefficient for different types of catchments is as follows:

Roof top- 0.75-0.95

Paved area- 0.50-0.85

Bare ground – 0.10-0.20

Green area – 0.05-0.10

Ground water recharge structures: Following are the recharge structures used for this purpose-

(a) Recharge pits

(b) Trenches

(c) Abandoned dug well

(d) Abandoned or running hand pump

(d) Abandoned tube well

(e) Recharge well

(f) Recharge shaft

(g) Lateral shaft with bore well

WATERSHED MANAGEMENT

The term watershed is defined as the land area from which water drains under gravity to a common drainage channel. Water shed is a delineated area with a well defined topographic boundry and one water outlet. In simple forms, a watershed refers to the area from where water to a particular drainage system, like a river or stream, comes from. Thus watershed may range from a few square kilometers to thousand square kilometers in size. In the watershed the hydrological conditions are such that water becomes concentrated within a particular location like a river or reservoir, by which the watershed is drained. A watershed comprises of complex interactions of soil, landform, vegetation, land use activities and water. Similarly people and animal are an integral part of a watershed having mutual impact on each other.

Watershed management is one of the critical factors for improving agricultural production. The natural resource base on which existence of living beings depends are under degradation. Loss of vegetal cover, followed by soil degradation through erosion, has resulted in lands lacking in water as well as soil nutrients. The rational utilization of land and water resources for optimum production causing minimum damage to the natural resourcesis known as watershed management. Rehabilitation of watershed through proper land use by adopting conservation strategies for minimum soil erosion so as to ensure good productivity of land for farmers is one of the important objectives of the watershed management.

The watershed management approach was included in the 5th Five Year Plan of India with a number of programmes for it and accordingly a national policy was developed. The practices of conservation and development of land and water are taken up with respect to their suitability for the benefit of the people and also sustainability pf the resources. The different measures taken up in this regard are as under:

(i) Water harvesting

(ii) Afforestation and agro-forestry

(iii) Measures to reduce soil erosion and runoff loses.

(iv) Mining and quarrying in a scientific manner

(v) Public participation.

IV. RESETTLEMENT AND REHABILITATION OF PEOPLE: PROBLEMS AND CONCERNS

Over the last 5 decades since independence high dams in India become more and more socially unjust, economically non-viable and environmentally disastrous. These dams instead of being labor intensive, they become capital intensive, leading to economic concentration, they are unable to provide protection to overall catchments of river valley basins and ultimately they have been cause of land degradation and deforestation. Today, more than 5.5 million hectares lies unutilized out of total 32.5 million hectares of potential created by high dams. The saddest story in the history of high dam development is the problem of dam oustees. The model of development adopted by most of the third world countries has necessitated the undertaking of large developmental projects like dam and defense installation etc. Land, therefore, is acquired from people who are the real owners of that land in proposed project area.

Therefore, displacement is a process that could encompass a series of phenomenon. No one term defines this process clearly. However, words like eviction, migration, deprivation of livelihood, deprivation of identity are used to approximate the phenomenon of displacement. Hence displacement is gradual, apparently invisible process, which ultimately leads to violation of peoples right to live, livelihood and identity.

Today over 2.5 million people live on every million hectare of land and by the end of century this will grow to 3 million. Therefore, as time goes by, the displacement problem will only grow. As forest land is protected there will be little land available for the resettlement. Mounting landlessness and distribution of the growing population is causing involuntary and accelerating migration of persons from the remote hills, tribal and other rural areas towards the urban centers. Due to this dams have become synonymous with submergence and displacement. As per the World Bank report, 11 irrigation dams constructed between 1978-88 and a total of 75,000 families were displaced.

Any policy of rehabilitation must attempt to recreate for displaced persons, a situation allowing them to improve or to restore the standard of living they had prior to displacement. Resettlement and rehabilitation must be treated as a whole community approach to take care of needs of all. We lack adequate policy guidelines and legal frame work. According to an estimate the number of persons directly or indirectly affected by irrigation projects in last 4 decades is about 20 million. During VIII five year plan 16.96 million people were displaced out of whom 8.14 lakh were tribal. Tribal communities have an ethos and their life is based upon their natural resources. Apart from depriving them of their land, livelihood and resources, displacement has other psychological and socio-cultural consequence which must be taken into account.

Displacement results in dismantling production system, desecrating ancestral sacred zones, graves and temples, weakening cultural system of self management and control The consequences are specially severe to woman. They loose access to the fuel, fodder that they collected for their home from common lands.

Sardar Sarovr dam in Gujrat will inundate 245 villages and displace 90,000 people. More than half of the Sardar Sarovar oustees will be tribal groups. – if they are not provided with suitable land, they cannot have a sustained life because they have no skills to face the competitive society. Narmada Sagar dam uptream of Sardar Sarovar in MP have displaced approximately 2.15 lacs people in catchments. There appears no socially justifiable rehabilitation due to land paucity.

Tehri dam has already inundated Tehri town and 25 villagers affecting 80,000 inhabitants. So far only small number of families have been resettled by clearing 2726 acres of sal forest in Bhaniyavala in eastern Doon.

V. ENVIRONMENTAL ETHICS

Environmental ethics is the ethical relationship between human beings and the environment in which they live. There are many ethical decisions that human beings make with respect to environment. For example:

(i) Should the human beings continue to clear cut the forests for the sake of their consumption.

(ii) Should they continue to make gasoloine powered vehicles, depleting fossil fuels while the technology exists to create zero emission vehicles.

(iii) What environmental obligations do they need to keep for future generations.

(iv) Is it right for humans to knowingly cause the extinction of a species for the convenience of humanity.

Environmental ethics thus consists of the study of normative issues and principles relating to human interactions with the natural environment and to their context and consequences. It forms a crucial area of applied ethics- crucial for the guidance of the individuals, corporations and governments in determining the principles affecting their policies, lifestyles and actions across the entire range of environmental issues.

The scope of environmental ethics is as extensive as its sphere, the realm of actions, policies which impact on natural environment. The spatial extent of its scope is at least as extensive as the biosphere. The temporal scope of environmental ethics extends for as long as human action can exercise any kind of impact, and as long as some thing of value remains on which significance impact can be made. Environmental ethics also studies the past in order to discover the traditions which often underlie current values, and which often turn out to supply limits to possible changes in ethical attitudes or resources for such change.

Proposals for an extension of ethics to cover all the species of the living systems of the earth first emerged in the 20th Century, in Aldo Leopold's A Sand County Almanac . In Leopolds land ethics, the land is the community of the interdependent species of the planet, plus the other components of their ecosystems. It was Leopold's proposal that a thing is right when it tends to promote the integrity, stability and beauty of biotic community. It is wrong when it tends otherwise.

Some think that environmental policies should be evaluated solely on the basis of how they affect humans. This entails a human centered environmental ethics. This ethic could lead to substantial agreement with environmentalists about policy. This would depend on the facts about the effects on humans of changes to the natural environment. This ethics treats only humans as morally considerable. There is another view which counts not only humans but non human animals also. But the life centered ethics includes more than humans and other animals, it includes plants, algae, all sorts of life forms existing on this planet and some times the ecosystems and the whole biosphere. A life centered ethics thus requires that in deciding how we should act and we need to take account of the impacts of our actions on every living thing affected by them.

VI. GREEN HOUSE EFFECT, GLOBAL WARMING AND CLIMATE CHANGE

The term green house effect was coined by Jean Fourier more than 100 years ago. The earth's surface and atmosphere receive energy from the sun. This energy comes in the form of radiation mainly in the visible region, l=400-700nm. Much of UV light (l< 400nm) from the sun is filtered out in the stratosphere and warms the air there rather than at surface of earth. Besides visible light, we receive some infra red (IR) light in the region 800-4000nm. Of the total incoming light that impinges upon the earth, about 50% reaches the surfaces and is absorbed by it. About 20% of the incoming light is absorbed by gases in the atmosphere, such as, UV light by ozone in stratosphere and IR by CO_2 and water vapor in air. The remaining 30% is reflected back in to space by clouds, ice, snow and other reflecting bodies without being absorbed.

Like any other warm body, the earth emits energy, the amount of the energy that the planet absorbs and the amount that it releases must be equal if the temperature is to remain constant at earth. The emitted energy is neither visible nor UV, it is IR light having wavelength in the range of 4000-5000nm which is called thermal infra red region.

Some gases in the air such as CO_2, temporarily absorb thermal infra red light and, therefore, not all IR emitted from the earth's surface and atmosphere escape directly in to space. Shortly after its absorption by CO_2 molecules, this thermal IR light is remitted in all directions. Thus some thermal IR is redirected back towards the earth's surface and is reabsorbed and consequently further heats to both, the surface and the air. This phenomenon of redirection of thermal IR towards the earth is called the ***green house effect***. Thus, the green house effect is trapping of heat caused by gases such as carbon di-oxide and water vapor which are transparent to incoming solar radiations but re-emit the infra red radiations from earth's surface. This is responsible for keeping the temperature of earth's surface at +15° C rather than -15° C (if there would had been no atmosphere).

The phenomenon that worries us is that, increasing the concentration of trace gases in atmosphere, that absorb thermal IR, would result in the redirection of even more of the out going thermal IR energy and would increase the average temperature of earth surface beyond+15° C. Most of the heat is absorbed by CO_2 layer and waters vapors in the atmosphere, which adds to the heat that is already present. CO_2 increases the earth temperature by 50% while CFCs are responsible for another 20% increase and there are enough CFCs up there to last in 120 years.

The heat trap provided by atmospheric CO_2 probably helped to create the conditions necessary for the evolution of life and the greening of earth. Compared to moderately warm planet, Mars, with too little CO_2 in its atmosphere, is frozen cold and Venus with too much is a dry furnace. The excess CO_2 to some extent is absorbed by the oceans. But with the industrialization of West and increased consumption of energy, CO_2 was released into atmosphere at a faster rate than the capacity of oceans to absorb it. Thus its concentration increased. According to some estimates CO_2 in air may have risen by 30% since the middle of 19th century. It may even be doubled by 2030 A.D.

GLOBAL WARMING

The global temperature has been observed to increase ever since systematic recording of this temperature has been initiated. The world temperature has increased by 0.7°C from 1860 to 1990AD. Where as temperature of tropical ocean has risen by 0.5°C for the last 30 to 50 years. It is estimated that the earth's mean temperature will rise between 1.5 t0 5.5 °C by 2050 if input of green house gases continue to rise at present rate.

CLIMATE CHANGE

The global change in temperature will not be uniform everywhere and will fluctuate in different regions. The places in higher latitudes will be warmed up more during late autumn and winter than the places in tropics. Poles may experience 2 to 3 times more warming than the global average while warming in the tropics may be only 50 to 100% of the average. The increased warming at poles will reduce the thermal gradient between the equator and high latitude regions decreasing the energy available to the heat engine that drives the global weather machine. This will disturb the global pattern of winds and ocean currents as well as timings and distribution of rainfall. Disturbed rainfall will result in some areas becoming wetter and the others drier.

Is the World's Climate Really Changing ? Many Scientists believe so. The UN's Intergovernmental Penal on Climate Change (IPCC) summarizes the work of 2,000 of the world's top most climate experts. Its reports say that the world is definitely getting warmer. The IPCC says the average global surface temperature has risen by about 0.6 degrees Celsius since 1900, with much of that rise coming in the 1990s- probably the warmest decade in 1,000 years. The IPCC also found that the snow cover since the late 1960s has decreased by 10% and lakes and rivers in the Northern Hemisphere are frozen over about two weeks less each year than they were then. Mountain glaciers in non-polar regions have also been in noticeble retreat in the 20th centuary, and the average global sea level has risen between 0.1 and 0.2 meters since 1900.

ACID RAINS

In the early 1970s it was noticed that lake without any known source of acid (e.g. mine seepage) in Canada, U.S. and Scandinavia were becoming increasingly acidic and that the fish populations of these lakes were being depleted. Acid from the sky was the only explanation, and indeed, monitoring demonstrated that the acidity of rainfall was well above the natural acidity of rain. This became known as **acid rain**. In fact, the term **acid deposition** is more accurate, as acid-forming materials may be deposited from the air in the form of snow, sleet, fog as well as in the form of rain. Acid rain should have been no surprise. For over a century we have been burning large quantities of oil and coat and smelting ore. Coal, and to a lesser extent oil contain sulfur. In the presence of oxygen and high combustion temperatures, sulfur compounds are oxidized to become sulfur oxides (SO_x). Sulfur dioxide is itself a poison, but it can also react with ozone, hydrogen peroxide and water vapor in the atmosphere to form sulfuric acid (H_2SO_4). Combustion at high temperatures in power plants and smelters also create oxides of nitrogen, mostly as atmospheric nitrogen combines with oxygen. Although nitric oxide (NO) is not very harmful as it does not readily dissolve, nitric oxide can combine with oxygen to form nitrogen dioxide :

$$2NO + O_2 = 2NO_2$$

Nitrogen dioxide is similar to sulfur dioxide; through various reactions with substances in the atmosphere, nitrogen dioxide is converted into nitric acid (HNO_3). Acid deposition has recently begun to cause obvious damage to the animals and plants in susceptible waters and soils where rain falls most heavily. Intensification of the acid rain problem from the 1950s through the 1970s in the US was observed. In eastern Canada, acid rain has rendered hundreds of lake fish free. Crops and other plants have also been affected. Sweden is reported to have around 3000 dead lakes, lakes without fish or frogs and where algae are the only visible from of life. All living things have an optimal pH level and limits. Departure from near optimal pH means sub-optimal reproduction, growth and survival. Acid rain changes the pH of soil and lakes while acidification can also cause toxic metals (e.g. aluminium and mercury) to be leached from soils and sediments. In addition, the detrimental effects of acid rain on forests was evident in Northern Europe in the 1980s.

OZONE (O3)

In the atmosphere, ozone is largely found in the stratosphere. Ozone absorbs most of the UV radiation of the sun and thus acts as a shield, protecting the living organisms on earth from health hazards. The depletion of this O3 layer by human activities may have serious implications and this has become a subject of much concern over the last few years. Ozone is also formed in the atmosphere through chemical reactions involving certain pollutants (SO2, NO2, aldehydes) on absorption of UV-radiations. The atmospheric ozone is now being regarded as potential danger to human health and crop growth.

The temperature decreases with increasing altitude in the troposphere (8 to16 km. from earth surface), while it increase with increasing altitude in the stratosphere (above 16 km. up to 50 km). This rise in temperature in stratosphere is caused by the ozone layer. The ozone layer has two important and interrelated effects. Firstly, it absorbs UV light and thus protects all life on earth from harmful effects of radiation. Second, by absorbing the UV radiation the ozone layer heats the stratosphere, causing temperature inversion which limits the vertical mixing of pollutants, thereby causing the dispersal of pollutants over the large areas near the earth surface. It is why a dense clouds usually hang over the atmosphere in highly industrialized areas causing several unpleasant effects. The wastes spread horizontally relatively fast (than slow mixing vertically), reaching all longitudes of the world in about a week and all latitudes within months. There is, therefore, very little that a country can do protect the ozone layer above it. The ozone problem is thus global in scope. In spite of slow vertical mixing, some of the pollutants (CFCs) enter the stratosphere and remain there for years until they are converted to other products or are transported back to the stratosphere. The stratosphere could be regarded as a sink, but unfortunately, these pollutants (CFEs) react with the ozone and deplete it.

The ozone near the earth's surface in the troposphere creates pollution problems. Ozone and other oxidants such as peroxy acetyl nitrate (PAN) and hydrogen peroxide are formed by light dependent reactions between NO2 and hydrocarbons. Ozone may also be formed under UV-radiation effect. These pollutants cause photochemical smog. Increase in O3 concentration near the earth's surface reduces crop yields significantly. It also has adverse effect on human health. Thus, while higher levels of ozone in the atmosphere protects us, it is harmful when its comes in direct contact with us and plants at earth's surface.

In plants, ozone enters through stomata. It produces visible damage to leaves, and thus a decrease in yield and quality of plant products. At 0.02 ppm it damages tobacco, tomato, bean, pine and other plants.

Ozone alone and in combination with other pollutants like SO_2 and NOx is causing crop losses of over 50% in several European countries. In Denmark, O_2 affects potato, clover, spinach, alfalfa etc. In limited pockets O3 concentration can be potentially harmful. For instance, in West Germany 100-250 mg/m3 O3 is not uncommon. In the Netherlands, O3 concentration was high enough to reduce yields of potato, beans and poplars. In several locations in U.K., O3 concentration exceeded 400 mg/m3 in 1976.

Ozone also reacts with many fibers especially cotton, nylon and polyester, and dyes. The extent of damage appears to be affected by light and humidity. O3 hardens rubber. A concentration of 0.3ppm causes nose and throat irritation, 1-3 ppm causes extreme fatigue while a concentration of 9.0 ppm causes severe pulmonary edema.

O3 is intimately connected with the life-sustaining process. Any depletion of ozone would, therefore, have catastrophic effects on life systems of the earth. Over the last few years, it could be realized that the O3 concentration of earth's atmosphere is thinning out. What has caused this depletion?

OZONE LAYER DEPLETION

In stratosphere, ozone absorbs UV rays causing the temperature inversion. This temperature inversion limits the vertical mixing of pollutants. However, in spite of this slow vertical mixing, some pollutants enter the stratosphere and remain there for years until they react with ozone and converted to other products. These pollutants thus deplete ozone in the stratosphere. Major pollutants responsible for this depletion are chlorofluorocarbons (CFCs), nitrogen oxides (coming from fertilizers) and hydrocarbons. CFCs are widely used as coolants in air conditioners and refrigerators, cleaning solvents, aerosol propellants and in foam insulation. CFC is also used in fire extinguishing equipment. They escape as aerosol in the stratosphere. Jet engines, motor vehicle, nitrogen fertilizers and other industrial activities are responsible for emission of CFCs, NOx etc. The supersonics air crafts flying at stratosphere heights cause major disturbances in O3 levels. The threat to ozone is mainly from CFCs, which are known to deplete O3 by 14% at the current emission rate. On the other hand NOx would reduce O3 by 3.5%. The nitrogen fertilizers release nitrous oxide during de-nitrification. Depletion of O3 would lead to serious temperature changes on the earth and consequent damage to life support system.

Depletion of ozone in stratosphere causes direct as well as indirect harmful effects. Since the temperature rise in stratosphere is due to heat absorption by ozone, the reduction in ozone would lead to temperature changes and rainfall failures on earth. Moreover one per cent reduction in O3 increase UV radiation on earth by 2%. A series of harmful effects are caused by an increase in UV radiation. Cancer is the best-established threat to man. When the O3 layer becomes thinner or has holes, it causes cancers, especially relating to skin like melanoma. A 10% decrease in stratospheric ozone appears likely to lead a 20-30% increase in skin cancer. The other disorders are cataracts, destruction of aquatic life and vegetation and loss of immunity. Nearly 6,000 people die of such cancers in USA each year. A 7% increase in such case has been reported from Australia and New Zealand.

Protecting ozone layer: Scientists throughout the world have shown a great concern over the thinning of ozone layer. This has led to a series of conferences on this issue. The first global conference on the depletion of ozone layer was held in Vienna (Austria) in 1985, the year, scientists discovered hole in South Pole. British team discovered a hole in ozone layer as large as that of the United States. This was followed by Montreal Protocol in 1987, which called for a 50% cut in the use of CFCs by 1998 reducing it to the level of 1986. Many countries including India did not sign the Protocol. India did not see any rationale as its release of CFCs is just 6,000 tones a year, equivalent to one and a half day's of world total. In our country per capita consumption of CFC is 0.02 kg. against 1 Kg. of developed world. CFCs are mainly the problem of developed world, as 95% of CFCs is released by European counties, U. S. A, USSR and Japan. USA alone releases 37% CFCs (producing CFC worth 2 billion dollar). The three-day international conference on Saving the ozone layer was organized jointly in London in March 1989 by the British Government and the UNEP. This conference highlighted the global problem created by the developed world, which in turn, is trying to dictate its terms to the developing countries for CFC pollution. The Montreal Protocol was initially signed by 31 countries.

There was held another international conference on ozone at Helsinki in May 1989 to revise the Montreal protocol. As many as 80 nations agreed to have a total ban on chemicals that cause ozone depletion by 2000 A.D. However, the conference backed away from a plan put forward by Dr. Mostafa Tolba, Executive Director, UNEP to set up an international climate fund. While the developing countries preferred to have the fund, the developed ones, including Japan, USA and UK rejected the plan. The agreement for CFC elimination by 2000 A. D. was needed as a major step towards environmental protection.

VII. NUCLEAR ACCIDENTS

During the night of April 25-26, 1986, the world's worst nuclear accident to date occurred at Chernobyl in Ukraine. The Chernobyl nuclear accident led to a severe release of fission production with high power of hundreds of millions of curies per hour which lasted during ten days. According to the Soviet reports to the IAEA in Vienna in 1986, the total radioactivity released was in the order of 100MCI. Due to the carelessness of the workers who were conducting an experiment with the nuclear reactor, an explosion occurred and blew the protective slab off the top of the reactor vessel. Lumps of radioactive material were ejected from the reactor and deposited within one kilometer of the plant. The main plume of the radioactive dust and gas was sent into the atmosphere, which was rich in fission products and contained iodine-131 and caesium-137, both of which can be readily absorbed by the living tissues. In this accident 31 people died and over 200 people sustained serious injuries through exposure to over 2000 times the normal annual dose from background levels of radiations.

VIII. WASTEL AND RECLAMATION

The unproductive soil usually considered wasteland. If the soil is suffering from hardening of the surface due to excessive consolidation, the only way to restore its health is to excavate and dig upto a depth of about 50 cm, so that water is able to infiltrate through openings in the dug up material. The soil is then seeded weather permitting, with application of requisite amounts of fertilizer including organic manure and leguminous plants. In some cases the seeded soil may have to be covered with a mulch of fibrous material such as chopped straw, shredded bark and wood pulp to protect the surface from rains and wind. In only extreme situations, a plastic or latex stabilizer (Table) is sprayed after seeding. A simple and pragmatic method is to mix the grass seeds with a small proportion of rapidly growing steamy plants, such as sorghum, which acts as a nurse crop (Bradshaw and Chadwick, 1980).

Table :Mulches and stabilizers suitable for restoring derelict lands

Material	Rate (t/ha)	Persis-tence	Stabili -zation	Soil-water Retention	Nutrients	Toxicity
Mulches						
Wood Shavings	4	xx	x	x	-	-
Shredded bark	4	xxx	x	xx	-	-
Jute netting		xx	xxx	x	-	-
Corncobs	10	xxx	x	x	-	-
Hay	3	x	x	x	x	-
Straw	3	xx	x	x	-	-
Stabilizers						
Wood cellulose fibre	1-2	xx	xxx	x	-	-
Sewage sludge	2-4	x	xx	-	-	-
Asphalt (1:1)	0.75	x	xx	x	-	x
Latex	0.2	x	xx	-	-	xx
Alginate/Colloidal carbonate	0.2	x	xx	-	-	-

(xxx high; xx moderate ; x low; - nil (Bradshaw and Chadwick, 1980)

These efforts must be supplemented by construction of a very efficient system of drainage to prevent access of runoff flow into the seeded ground. Since the derelict lands are devoid of nutrients, ample doses of fertilizers with proportionate amounts of leguminous plants are added to the soil. Legume with *Rhizobium* bacteria attached to them would provide nitrogen to the recuperating soil Inoculation of seeds with *Rhizobium* culture would expedite the process of soil recovery. Soil suffering from toxicity requires special treatment. Acidity can be countered with application of lime, and the salinity can be leached out by excess irrigation or by planting species that ensure eradication of this malaise.

Severely degrade lands are best retired permanently for plantation of fuel and fodder trees, or species that are suited to the local climatic and soil conditions. The usual method is to transplant 1-3 year old seedlings raised in nurseries (or less than a year-old grown in small fibre tubes or polythene containers). In case of certain plant, chipped branches or seed bearing shoots can be spread on the soil and allowed to sprout. Needless to state, the application of fertilizers to supply nutrients and also irrigation would be necessary to ensure easy or rapid growth of seeds, seedlings and sproutable chips.

IX. CONSUMERISM AND WASTE PRODUCTS

The traditional cultural values are degenerating under the influence of corporate politics, the commercialization of culture and the mass media. The people have been inundated by an unending parade of commodities and fabricated television spectacles that keeps it preoccupied with the ideals and the values of consumerism. Consumerism is the myth that the individual will be gratified and integrated by consuming. This will lead to affluenza. Jessie O'Neill defines it as a dysfunctional relationship with wealth or money. Symptoms of affluenza include a love of shopping. It is plague of materialism and over consumption that is so pervasive that it characterizes our modern society.

United States leads the world in per capita waste generation. This is a society wide problem. Such trends lead to huge waste production. The sorting of waste material give an idea about the composition of solid waste.

- Paper (41% recovery) can be repulped and re-procssed into recyclic paper, cardboards and other paper products, shredded and composted.
- Glass (26% recovery) can be crushed, remitted and made into containers or used as substitute for gravel in construction material such as concrete and asphalt.
- Some forms of plastic (5.4% recovery) can be re-melted and fabricated into carpet fiber outdoor wearing apparel, irrigation drainage tiles.
- Metals can be re-melted and re-fabricated aluminium (38% recovery) from scrap, aluminium saves up to 90% of the energy required to make aluminium from ore.
- Food wastes and yard waste (39% recovery) can be composted to produce a humus soil conditioner.
- Textiles (12% recovery) reuse or can be shredded and used to strengthen recycled paper products.
- old times (19% recovery) can be melted or shredded and incorporated into highway asphalt.

Recycling of waste is probably the most direct and obvious way most people can become involved in environmental issues. If you recycle, you save some natural resources from being used and you prevent landfills from landfulls.

CHAPTER-XIV

Acts and Laws

The scientific community had never been unaware of the fact that there exists a vital link between environment and all forms of life on this earth. The balance between different environmental components is the pre-requisite for a healthy existence, and consequently, the preservation of the essential ingredients of life does require a stable ecological balance. The environmental issues have cut across the barriers of languages, culture, religions and national boundaries and thus are being looked upon from a global point of view rather than national. In its first conference on the Human Environment in June 1972, the United Nations showed a high concern on the fast decreasing resources and degrading environment. The General Assembly of the United Nations passed a Resolution on 15th December 1972 giving emphasis on the need for an active cooperation among the States in the field of human environment, and designated 5th June as the World Environment Day. The recommendations made by the General Assembly included short and long term plans at regional, national and international levels with particular reference to the environment and development. The Principal 1 of the Stockholm declaration, declared:

(a) Man has the fundamental right to freedom, equality and adequate conditions of life, in an environment of quality that permits a life of dignity and well being, and

(b) Man bears a solemn responsibility to protect and improve the environment for present and future generations.

The constitution of India has specific provisions for the protection of environment which were incorporated in the constitution after the Stockholm conference on Human and

Environment. An amendment in the Constitution for the first time inserted relevant provisions in part IV (Directive Principles) and part IV A (Fundamental Duties) of the constitution. The Article 47 provides for the raising of the level of nutrition, living standards and improvement of public health as the primary duties of the state. The Article 48 provides the state to organize agriculture and animal husbandry on modern and scientific lines and take steps for preservation and improving the breeds. The state shall endeavor to protect and improve the environment and to safeguard the forests and wildlife of the country. Article 51 specifically deals with the fundamental duties with respect to environment . It states : It shall be the duty of every citizen of India to protect and improve the natural environment including the forests, lakes, rivers and wildlife and to have compassion for living creatures.

Thus, along with the developmental process, a great emphasis is also laid on taking preventive and control measures to combat the problems related with environment.

Giving the utmost of the importance to the environmental conservation, Government of India constituted a high power committee in 1980 and Mr. N. D. Tewari was made the chairman of the committee. This committee was assigned functions to recommend legislative measures, formulation of environmental guidelines, creating environmental awareness at different levels and propose the administrative setup so as the protection of environment is fully ensured. This committee submitted its report in 1980 and in its review of the environmental legislation, noted the following shortcomings in the existing laws:

(i) Most of such laws were outdated.

(ii) The laws lacked the statements of explicit policy objectives.

(iii) The laws lack adequate provisions for helping the machinery for their implementation.

(iv) The laws were mutually inconsistent, and

(v) There was no procedure for reviewing the efficacy of those laws.

On the basis of a thorough analysis of existing laws, the committee made the following recommendations:

(i) A comprehensive review and reformation of some Central and State Acts.

(ii) New legislation for areas of action not covered by existing laws.

(iii) Introduction of environment protection in the concurrent list of the seventh schedule.

It is now being realized by each and every country of the world that effective laws should be enacted and rules should be made for the protection and preservation of environment and there have been a series of International Conventions for preservation and protection of the environment. The United Nations General Assembly on 29th October 1982 adopted "the world charter for nature", which declares that:

(a) Mankind is a part of nature and life depends on the uninterrupted functioning of natural systems which ensure the supply of energy and nutrients.

(b) Civilization is rooted in nature, which has shaped human culture and influenced all artistic and scientific achievements, and living in harmony with nature gives man the best opportunities for the development of his creativity, and for rest and recreation.

The important acts which have been passed and enacted by the Parliament of India are as under:

(i) The Wild Life (Protection) Act, 1972.

(ii) The Forest Conservation Act, 1980.

(iii) The Environment (Protection) Act, 1986.

(iv) The Water (Prevention and Control of Pollution) Act, 1974

(v) The Air (Prevention and Control of Pollution) Act, 1988

(vi) The Biodiversity Act, 2002.

(vii) The national environmental tribunal Act, 1995

All the laws pertaining to conservation and protection of environment have been grouped in to the following two heads by Prof. Chhatrapati Singh:

(a) Protective legislation which are concern with the human beings and non-human beings, providing protection to human beings and others from the unjust harm caused by other human beings.

(b) Planning legislation related with production and distribution aiming to regulate the use of resources in a sustainable manner so as to maintain an ecological equilibrium and; equitable distribution of resources among those who share it.

The important acts have been very briefly discussed here below:

THE ENVIRONMENT (PROTECTION) ACT, 1986

No. 29 of 1986 (23rd May, 1986)

An Act to provide protection and improvement of environment and for matters connected therewith.

CHAPTER 1. PRILIMINARY

1. Short title, extent and commencement.

2. Definitions.

(a) Environment includes water, air and land and the interrelationship which exists among and between water, air and land and human beings., other living creatures, plants, micro-organisms and property.

(e) hazardous substances means any substance or preparation which by reason of its chemical or physico-chemical properties or handling, is liable to cause harm to human beings, other living creatures, plants, micro-organisms, property or the environment.

CHAPTER II. GENERAL POWERS OF THE CENTRAL GOVERNMENT.

3. Power of Central Government to make measures to protect and improve environment.

4. Appointment of officers and their powers and functions.

1. Power to give directions

2. Rules to regulate environmental pollution.

CHAPTER III. PREVENTION, CONTROL AND ABATEMENT OF ENVIRONMENTAL POLLUTION.

3. Persons carrying on industry operation, etc. not to allow emissions or discharge of environmental pollutants in excess of the standards.

4. Persons handling hazardous substances to comply with procedural safe guards.

5. Furnishing of informations to authorities and agemcies in certain cases.

6. Power of entry and inspection.

7. Power to take sample and procedure to be followed in connection therewith.

8. Environmental laboratories.

9. Government analysts

10. Reports of Government analysts.

11. Penalty for contravention of the provisions of the Act and the rules, orders and directions.

12. Offences by companies.

13. Offences by Government Departments.

CHAPTER IV. MISCELLANEOUS

14. Protection of action taken in good faith.

15. Cognizance of offences.

16. Information, reports and returns.

17. Members, officers and employees of the authority constituted under section 3 to be public servants.

18. Bar of jurisdiction

19. Power to delegate.

20. Effect of other laws.

21. Power to make rules.

Rules made under this Act to be laid before the Parliament.

THE AIR (PREVENTION AND CONTROL OF POLLUTION) ACT, 1981

Act No. 14 of 1981 929th March, 1981)

An Act to provide for the prevention, control and abatement of air pollution, for the establishment, with a view to carrying out the aforesaid purposes, of Boards, for conferring on and assigning to such Boards powers and functions thereto and for matters connected therewith.

Whereas decisions were taken at the United Nations Conference on Human Environment held in Stockholm in June 1972, in which India participated, to take appropriate steps for the preservation of the quality of air and control of air pollution;

And whereas it is considered necessary to implement the decisions aforesaid insofar as they relate to preservation of the quality of air and control of air pollution;

Be it enacted by Parliament in the Thirty second year of the Republic of India as follows-

CHAPTER I. PRILIMINARY

1. Short title, extent and commencement.

2. Definitions-

(a) air pollutant means any solid, liquid or gaseous substances (including noise) present in the atmosphere in such concentration as may be or tend to be injurious to human beings or other living creatures or plants or property or environment;

(b) air pollution means the presence in the atmosphere of any air pollutant.

CHAPTER III. CENTRAL AND STATE BOARDS FOR THE PREVENTION AND CONTROL OF AIR POLLUTION.

3. Central Pollution Control Board.

4. State Pollution Control Boards constituted under Section 4 of Act 6 of 1974 to be State Boards under this Act.

5. Constitution of State Boards.

6. Central Board to exercise the powers and perform the functions of State Board in Union Territories.

7. Terms and conditions of services of members.

8. Disqualifications.

9. Vacation of seats by members.

10. Meetings of Board.

11. Constitution of committees.

12. Temporary association of persons with Board for particular purposes.

13. Vacancy in Board not to invalidate acts or proceedings.

14. Member-secretary and officers and other employees of State Boards.

15. Delegation of powers.

CHAPTER III. POWERS AND FUNCTIONS OF BOARDS

16. Functions of Central Board.

17. Functions of State Boards.

18. Power to give directions.

CHAPTER IV. PREVENTION AND CONTROL OF AIR POLLUTION

19. Power to declare air pollution control areas.

1. Power to give instructions for ensuring standards for emissions from automobiles.

2. Restriction on use of certain industrial plants.

3. Persons carrying on industry, etc. not to allow emission of air pollutants in excess of the standards laid down by the State Boards.

22-A. Power of Board to make application to court for restraining persons from causing air pollution.

4. Furnishing of information to State Board and other agencies in certain cases.

5. Power of entry and inspection.

6. Power to obtain information.

7. Power to take samples of air or emission and procedure to be followed in connection therewith.

8. Reports of results of analysis on samples taken under Section 26.

9. State air laboratory.

10. Analysts.

11. Reports of analysts.

12. Appeals.

31-A. Power to give directions.

CHAPTER V. FUNDS, ACCOUNTS AND AUDIT.

13. Contributions by Central Government.

14. Fund of Board.

15. Budget.

16. Annual report.

17. Accounts and audit.

CHAPTER VI. PENALTIES AND PROCEDURES

18. Failure to comply with the provisions of section 21 or section 22 or with the directions issued under section 31-A.

19. Penalties for certain acts.

20. Penalties for contravention of certain provisions of the Act.

21. Offences by companies.

22. Offences by Government Departments.

23. Protection of action taken in good faith.

24. Cognizance of offences.

25. Members, officers and employees of Board to be public servants.

26. Reports and returns.

27. Bar of jurisdiction.

CHAPTER VII. MISCELLANEOUS

28. Power of State Government to supersede State Board.

29. Special provisions in the case of suppression of the Central Board or State Boards constituted under the Water (Prevention and control of pollution) Act, 1974.

30. Dissolution of State Boards constituted under the Act.

31. Omitted by Act 47 of 1987.

32. Maintenance of register.

33. Effect of other laws.

34. Power of Central Government to make rules.

Power of State Government to make rules.

THE WATER (PREVENTION AND CONTROL OF POLLUTION) ACT, 1974

Act no. 6 of 1974 (23rd March, 1974)

An Act to provide for th prevention and control of water pollution and the maintaining or restoring of wholesomeness of water., for the establishment, with a view to carrying out the purposes aforesaid, of Boards for the prevention and control of water pollution, for conferring on and assigning to such Boards, powers and functions relating thereto and for matters connected therewith.

The different chapters and Sections and sub-sections of the Act have been briefly discussed below.

CHAPTER I : 1. SHORT TITLE, APPLICATION AND COMMENCEMENT- (1) This Act may be called the Water (Prevention and Control of Pollution) Act, 1974.

2. Definitions- Under this Section various terms used in the Act have been defined-

(e) pollution- means such contamination of water or such alteration of the physical, chemical or biological properties of water or such discharge of any sewage or trade effluent or of any other liquid, gaseous or solid substance into water (whether directly or indirectly) as may, or is likely to, create a nuisance or render such water harmful or injurious to public health or safety, or to domestic, commercial, industrial, agricultural or other legitimate uses, or to the life and health of animals or plants or of aquatic organisms.

(g) sewage effluent- means effluent from any sewerage system or sewage disposal works and includes sullage from open drains.

(gg) sewer-means any conduit pipe or channel, open or closed carrying sewage or trade effluent.

(j) Stream includes-

(i) river

(ii) watercourse (whether flowing or for the time being dry)

(iii) inland water (whether natural or artificial)

(iv) sub-terranean waters.

(v) sea or tidal waters to such extent or, as the case may be, to such point as the State Government may,. By notification in the official gazette, specify in this behalf.

CHAPTER II. THE CENTRAL AND STATE BOARDS FOR PROVISION AND CONTROL OF WATER POLLUTION.

3. Constitution of Central Board.

4. Constitution of State Boards.

5. Terms and conditions of service members.

6. Disqualifications.

7. Vacation of seats by members.

8. Meetings of Board.

9. Constitution of committees.

10. Temporary association of persons with Board for particular purposes.

1. Vacancy in Board not to invalidate acts or proceedings.

11-A. Delegation of powers to Chairman.

2. Member-Secretary and officers and other employees of Board.

CHAPTER III. JOINT BOARDS.

3. Constitution of Joint Boards.

4. Composition of Joint Boards.

5. Special provision relating to give of directions.

CHAPTER IV. POWERS AND FUNCTIONS OF BOARDS.

6. Functions of Central Board.

7. Functions of State Board.

8. Powers to give directions

CHAPTER V. PREVENTION AND CONTROL OF WATER POLLUTION

9. Power of State Government to restrict the application of the Act to certain areas.

10. Power to obtain information

11. Power to take samples of effluents and procedure to be followed in connection therewith.

12. Reports of the results of analysis on samples taken under Section 21.

13. Power of entry and inspection.

14. Prohibition on use of stream or well for disposal of polluting matter, etc.

15. Restriction on new outlets and new discharges.

16. Provision regarding existing discharge of sewage or trade effluent.

17. Refusal or withdrawal of consent by State Board.

18. Appeals.

19. Revisions.

20. Power of State Boards to carry out certain works.

21. Furnishing of information to State Boards and other agencies in certain cases.

22. Emergency measures in case of pollution of streams or well.

23. Power of Board to make application to courts for restraining apprehended pollution of water of streams or wells.

CHAPTER VI. FUNDS, ACCOUNTS AND AUDIT

24. Contributions by Central Government.

25. Contributions by State Government.

26. Fund of Central Board.

27. Fund of State Board.

37-A. Borrowing powers of Board.

28. Budget.

29. Annual report.

30. Accounts and audit.

CHAPTER VII. PENALTIES AND PROCEDURES.

31. Failure to comply with directions under sub-section (2) or sub-section (3) of Section 20, or orders issued under Clause © of sub-section (1) of Section 32 or directions issued under Sub-section (2) of Section 33 or Section 33-A.

32. Penalty for certain acts.

33. Penalty for contravention of provisions of Section 24.

34. Penalty for contravention of Section 25 or Section 26.

35. Enhanced penalty after previous conviction.

36. Publication of names of offenders.

37. Offences by companies.

38. Offences by Government Departments.

39. Cognizance of offences.

40. Members, officers and servants of Board to be public servants.

CHAPTER VIII. MISCELLANEOUS.

41. Central water laboratory.

42. State water laboratory.

43. Analysts.

44. Reports of analysts.

45. Local authorities to assist.

46. Compulsory acquisition of land for the State Boards.

47. Returns and reports.

48. Bar of jurisdiction.

49. Protection of action taken in good faith.

50. Overriding effect.

51. Power of Central Government to supersede the Central Board and Joint Boards.

52. Power of State Government to supersede State Board.

53. Power of Central Government to make rules.

54. Power of State Government to make rules.

THE WILDLIFE (PROTECTION) ACT, 1972

(NO. 53 OF 1972)

An Act to provide for the protection of wild animals, birds and plants and for that matters connected therewith or ancillary or incidental thereto.

The wildlife (Protection) Act, 1972 has seven chapters. Chapter 1 comprises of various definitions of the terms which have been used in this Act In the Act, Animal includes amphibians, birds, mammals and reptiles, and their young, and also includes, in the case of birds and reptiles, their eggs. Wild animal has been defined as any animal found wild in nature and includes any animal sp[edified in Schedules I, II, III, IV or V, wherever found. Wildlife includes any animal, bees, butterflies, crustacea, fish, and moths; and aquatic or land vegetation which forms part of any habitat. National Park means an area declared, whether under Sec.35 0r 38, or deemed, under sub-section (3) of Sec. 66, to be declared, as a National Park. Sanctuary means an area declared, whether under Section 18, 26 (A) or Sec. 38, or deemed, under sub section 66 to be declared, as a wildlife sanctuary. Trophy, means the whole or any part of any captive animal or wild animal, other than vermin, which has been kept or preserved by any means, whether artificial or natural, and includes (a) rugs, skin and specimens of such animals mounted in whole or in part through a process of taxidermy, and (b) antlers, horn, rhinoceros horn, hair, feather, nail, tooth, musk, eggs, and nests. Vermin means any animal specified in Schedule V. Zoo means an establishment, whether stationary or mobile, where captive animals are kept for exhibition to the public but does not include a circus and an establishment of a licensed dealer in captive animals.

Chapter 2 of the Act deals with the authorities to be appointed or constituted under the Act. Under the Section 3 of the Act the central government may appoint director and other officers. Under the Section 4 the State Government may , for the purpose of this Act appoint Chief Wildlife Warden, Wildlife warden and other officers. Under Sec. 5, powers delegated to the authorities have been mentioned. Constitution of wildlife advisory board has been provided in Sec. 6 of the Act.

Chapter III of the Act deals with the hunting of wild animals. Under Sec. 9 of the Act, No person shall hunt any wild animal specified in Schedules I,II, III and Iv except as provided under Sec. 11 and 12. Hunting of wild animals is permitted in certain cases under the Sec.11 While Sec. 12 grants permission /permit for special purposes.

Chapter IIIA of the Act deals with protection of specified plants.

Chapter IV deals with Sanctuaries(Sec. 18, 26 A and 38), National Parks(Sec. 35 and Sec. 38) and closed areas (Sec. 37)

Chapter IV A is for constitution of a Central Zoo Authority and provides the procedure for recognition of Zoos in the country. Under Section 38C functions of the authority has been enlisted and Section 38 H deals with the recognition of zoos.

Chapter V of the Act deals with trade or commerce in wild animals, animal articles and trophies. Under Section 39 of the Act every wild animal, other than vermin, which is hunted under section 11 or sub section 35 or kept or bred in captivity or hunted in contravention of any provision of this act or any rule pr order made there under, or found dead or killed by mistake will be the Government property. Animal articles, trophy or uncured trophy or meat derived from any wild animal; ivory imported into India and an article made from such ivory; vehicle, vessel, weapon, trap or tool that has been used for committing an offence and has been seized under the provisions of Act shall be the Government property.

Chapter V A of the Act deals with the prohibition of trade or commerce in trophies, animal articles etc. derived from certain animals. In this chapter, under Section 49 A, Scheduled animal means an animal specified for the time being in Schedule I or Part II of Schedule II. Scheduled animal articles means an article made from any scheduled animal and includes an article or object in which the whole or any part of such animal has been used, but does not include tail feathers of peacock, an article or trophy made therefrom and snake venom or its derivates.

Section 49 A (a) (c) (ii) in the amended Act provides only two months time to dispose of scheduled animal articles.

Section 49 B provides prohibition of dealing in trophies, animal articles etc. derived from Scheduled animals. Declarations are to be made by the dealers under Section 49 C of the Act.

Chapter VI of the Act deals with the prevention and declaration of offences. Power of entry, search, arrest and detention has been given in Section 50. Penalties have been fixed under Section 51 of this Act and, a person found guilty of an offence against this Act and on conviction be punishable with imprisonment for a term which may extend for three years or with fine which may extend to twenty five thousand rupees, or with both.

Chapter VII of the Act deals with miscellaneous matters e.g. Officers to be public servants (Sec. 59), protection of action taken in good faith (Sec. 60), reward to persons(Sec. 60A) and power to alter entries in Schedules(Sec. 61), power of Central Government to make rules(Sec. 63) and power of State Government to make rules (Sec. 64), and protection of rights of Scheduled tribes(Sec.65) etc.

THE FOREST (CONSERVATION) ACT, 1980

(Act No. 69 of 1980). It received the assent of the President on 27th December, 1980 and published in the Gazette of India (Extraordinary), Part II, Section1, dated 27th December 1980.

1. Short title, extent and commencement: (1) This Act may be called the Forest (Conservation) Act, 1980.

(2) It extends to the whole of India except the State of Jammu and Kashmir.

(3) It shall be deemed to have come into force on 25th day of October, 1980.

2. Restriction on the de-reservation of forests or use of forest land for non forest purpose: Notwithstanding anything contained in any other law for the time being in force in a state, no State Government or other authority shall make, except with the prior approval of the Central Government, any order directing-

(i) that any reserved forest (within the meaning of the expression "reserved forest" in any law for the time being in force in the State) or any portion thereof, shall cease to be reserved;

(ii) that any forest land or any portion thereof may be used for any non forest purpose.

(iii) that any forest land or any portion thereof may be assigned by way of lease or otherwise to any private person or to any authority, corporation, agency or any other organization not owned, managed or controlled by Government.

(iv) that any forest land or any portion thereof may be cleared of trees which have grown naturally in that land or portion, for the purpose of using it for reaforestation.

3. Constitution of advisory committee- The central Government may constitute a Committee consisting of such number of persons as it may deem fit to advise that Government with regard to-

(i) the grant of approval under section 2 and,

(ii) any other matter connected with the conservation of forests which may be referred to it by the Central Government.

3 A. Penalty for contravention of the provisions of the Act: Whoever contravenes or abets the contravention of any of the provisions of Section 2, shall be punishable with simple imprisonment for a period which may extend to fifteen days.

3B. offences by authorities and Government Departments: (1) Where any offence under this Act has been committed-

(a) by any department of Government, the head of the department, or

(b) by any authority, every person who at the time of the offence was committed, was directly in charge of, and was responsible to, the authority for the conduct of the business of the authority as well as the authority; shall be deemed to be guilty of the offence and shall be liable to be proceeded against and punished accordingly.

Provided that nothing contained in this sub-section shall render the head of the department or any person referred to in clause (b), liable to any punishment if he proves that the offence was committed without his knowledge or that he exercised all due diligence to prevent the commission of such offence.

(2) Not withstanding anything contained in sub-section(1), where an offence punishable under the Act has been committed by a department of government or any authority referred to in clause (b) of sub section (1) and it is proved that the offence has been committed with the consent or connivance of; or is attributable to any neglect on the part of any officer, other than the head of the department, or in the case of an authority, any person other than the persons referred to in clause (b) of sub-section (1) , such officer or persons shall be deemed to be guilty of that offence and shall be liable to be proceeded against and punished accordingly.

4. Power to make rules: (1) The Central Government may, by notification in the official Gazette, makes rules for carrying out the provisions of this Act.

(2) Every rule made under this Act shall be laid, as soon as may be after it is made, before each house of parliament, while it is in session, for a total period of thirty days which may be comprised in one session or in two or more successive sessions, and if , before the expiry of the session immediately following the session or the successive sessions aforesaid, both houses agree in making any modifications in the rule or both houses agree that the rule should not be made, the rule shall thereafter have effect only in such modified form or be of no effect, as the case may be; so, however,

that any such modification or annulment shall be without prejudice to the validity of anything previously done under that rule.

5. Repeal and saving: (1) The Forest (Conservation) Ordinance, 1980 is hereby replaced.

(2) Notwithstanding such repeal, anything done or any action taken under the provisions of the said ordinance shall be deemed to have been done under the corresponding provisions of this Act.

THE BIODIVERSITY ACT, 2002.

The Biological Diversity Act, 2002 was passed by the Indian Parliament and has received the ascent of the President on 5th February, 2003 (No. 18 of 2003).

CHAPTER 1: This is the preliminary one, which includes the following two clauses:

1. Short title, extent and commencement of the Act
2. Definitions.

CHAPTER 2: It is regarding the regulation of access to biological diversity and covers the following clauses:

3. Certain persons not to undertake Biodiversity related activities without approval of National Biodiversity Authority
4. Result of research not to be transferred to certain persons without the approval of National Biodiversity Authority.
5. Section 3 and 4 of the Act are not to apply to certain collaborative research activities.
6. Application for intellectual property rights (IPR) not to be made without approval of National Biodiversity Authority (NBA).
7. Prior intimation to State Biodiversity Boards for obtaining biological resources for certain purposes.

CHAPTER 3: This is regarding the National Biodiversity Authority and covers the following Clauses:

8. Establishment of National Biodiversity Authority.

9. Conditions of service of Chairperson and Members.

10. Chairperson to be Chief Executive of National Biodiversity Authority.

11. Removal of members

12. Meeting of National Biodiversity Authority.

13. Committees of National Biodiversity Authority.

14. Officers and employees of National Biodiversity Authority.

15. Authentication of orders and decisions of National Biodiversity Authority.

16. Delegation of Powers.

17. Expenses of National Biodiversity Authority to be defrayed out of the consolidated fund of India.

CHAPTER 4: This chapter deals with the functions and powers of the National Biodiversity Authority and covers following clause:

18. Functions and powers of National Biodiversity Authority

CHAPTER 5: It is regarding the approval by the National Biodiversity Authority and includes the following three clauses:

19. Approval by National Biodiversity Authority for undertaking certain activities.

20. Transfer of biological resource and knowledge.

21. Determination of equitable benefits sharing by National Biodiversity Authority.

CHAPTER 6: Deals with State Biodiversity Board and includes following clauses:

22. Establishment of State Biodiversity Board.

23. Functions of State Biodiversity Board.

24. Power of State Biodiversity Board to restrict certain activities violating the objectives of conservation, etc.

25. Provisions of sections 9 to 17 of the Act to apply with modifications to State Biodiversity Board

CHAPTER 7: Deals with finance, accounts and audit of National Biodiversity Authority. It has the following clauses:

26. Grants and loans by the Central Government.

27. Constitution of National Biodiversity Fund.

28. Annual Report of National Biodiversity Authority.

29. Budget, accounts and audit.

30. Annual report to be laid before the parliament.

CHAPTER 8: Finance, accounts and audit of state Biodiversity Board. It includes the following clauses:

31. Grant of money by State Government to State Biodiversity Board.

32. Constitution of State Biodiversity Board.

33. Annual Report of State Biodiversity Board.

34. Audit of accounts of State Biodiversity Board.

35. Annual report of State Biodiversity Board to be laid before State Legislation.

CHAPTER 9. It includes the duties of the Central and State Governments and has following clauses:

36. Central Government to develop national strategies and plans for conservation etc., of biological diversity.

37. Biodiversity heritage sites.

38. Power of Central Government to notify threatened species.

39. Power of Central Govt. to designate repositories.

40. Power of Central govt. to exempt certain biological resources.

CHAPTER 10: This chapter is regarding the constitution of biodiversity management committees and has following clause.

41. Constitution of Biodiversity Management Committee.

CHAPTER 11: Local biodiversity fund. It has the following clauses:

42. Grants to local biodiversity fund.
43. Constitution of local biodiversity fund.
44. Application of local biodiversity fund.
45. Annual report of local biodiversity management committees.
46. Audit of accounts of local biodiversity management committees.
47. Annual report of biodiversity management committee to be submitted to district magistrate.

CHAPTER 12: Deals with other various miscellaneous matters pertaining to biodiversity. It has following clauses:

48. National Biodiversity Authority to be bound by the directions given by central government.
49. Power of State Government to give directions.
50. Settlements of disputes between State Biodiversity Boards.
51. Members, officers etc. of National Biodiversity Authority and State Biodiversity Board deemed to be public servants.
52. Appeal.
53. Execution of determination or order.
54. Protection of action taken in good faith.

55. Penalties.

56. Penalties for contravention of directions or orders of Central Governments, State Governments, National Biodiversity Authority and State Biodiversity Boards.

57. Offences by companies.

58. Offences to be cognigible and non bailable.

59. Act to have effect in addition to other Acts

60. Power of central Government to give directions to State Governments.

61. Cognigence of offence.

62. Power of central Government to make rules.

63. Power of State Governments to make rules.

64. Power to make regulations.

65. Power to remove difficulties.

14.7 THE NATIONAL ENVIRONMENTAL TRIBUNAL ACT, 1995.

27 OF 1995 (17th June, 1995).

An Act to provide for strict liability for damages arising out of any accident occurring while handling any hazardous substance and for the establishment of a National Environmental Tribunal for effective and expeditious disposal of cases arising from such accidents, with a view to giving relief and compensation for damages to persons, property and environment and for matters connected therewith or incidental thereto.

CHAPTER I. PRILIMINARY.

1. Short title and commencement.

2. Definitions

CHAPTER II. COMPENSATION FOR DEATH OF OR INJURY TO A PERSON AND DAMAGE TO PROPERTY AND ENVIRONMENT.

3. Liability to pay compensation in certain cases on principal of fault.

4. Application for claim for compensation.

5. Procedures and powers of Tribunal.

6. Conditions as to making of interim orders.

7. Reduction of amount of relief paid under any other law.

CHAPTER III. ESTABLISHMENT OF NATIONAL ENVIRONMENTAL TRIBUNAL AND BENCHES THEREOF.

8. Establishment of National Environmental Tribunal.

9. Composition of Tribunal and Benches thereof.

10. Qualifications for appointment as Chairperson, Vice-Chairperson or Other members.

11. Vice-Chairperson to act as Chairperson or to discharge his functions in certain circumstances.

12. Term of office.

13. Resignation and removal.

14. Salaries and allowances and other terms and conditions of service of chairperson, vice-chairperson and other members.

15. Provision as to the holding of offices by Chairperson etc. on ceasing to be such Chairperson etc.

16. Financial and administrative powers of Chairperson.

17. Staff of Tribunal.

18. Distribution of business amongst the Benches.

CHAPTER IV. JURISDICTION AND PROCEEDINGS OF THE TRIBUNAL

19. Bar of jurisdiction.

20. Power of Chairperson to transfer cases from one bench to another.

21. Decisions to be taken by majority.
22. Deposits of amount payable for damage to environment.
23. Execution of award or order of Tribunal.
24. Appeals.

CHAPTER V. MISCELLANEOUS.

25. Penalty for failure to comply with orders of Tribunal.
22. Offences by companies.
23. Proceedings before the Tribunal to be judicial proceedings.
24. Members and Staff of Tribunal to be public servants.
25. Protection of action taken in good faith.
26. Act to have overriding effect.
27. Power to make rules.

ISSUE INVOLVED IN ENFORCEMENT OF ENVIRONMENT LEGISLATION

All over this world the exponential growth of people, production, power, place and pollutants are having their impact on the water, air and land cycles of nature to the detriment of mankind. Environmental protection is man centered. The problem is of total management of our water resources, of our excreta and wastewaters, of our air environment or our solid wastes and of our food to prevent impairment of health, to promote our efficiency and comfort and to safeguard the balance in natural ecosystems. Today, the world concern is on how to combat pollution and how to maintain the standard of human environment. In a developing country where finances are a great constraint the matter gets aggravated on account of lack of awareness. But non the less, India's efforts are no less than of any developing country. We are probably the only country having an article on environment in our constitution. India is signatory to practically all international conferences and conventions on Environment. India has passed several statutes for the protection of environment. The water (Prevent and Control of

Pollution) Act, 1974, Air (Prevention and Control) Act, 1976, Environment (Protection) Act, 1986. India is one of the few countries which has anti-pollution and protection of environment law, existing even in the first half own laws for regulating pollution. All these statutes are not faultless. Yet much more faulty in their implementation. The Tiwari committee that many existing laws primarily promotes development and resource utilization for specific economic benefit without analyzing the potential short and long term deleterious effects. Several Laws relating to the management of environmental resources do not clearly state social objectives, they aim to achieve. To prevent such flaws it has been suggested that since environ protection is of national importance, Parliament should initiate legislation to cover all aspects of environment throughout the country in an integrated manner. Efforts are now being made to make up for the gaps in laws by following principles of Public trust or Human Rights.

The 42nd constitutional amendment has inserted Article 48 A which states :

"The state shall endeavor to protect and improve the environment and to safeguard the forests and wildlife of the country".

Our constitution also lays down the fundamental duty to every citizen "to protect and improve the natural environment including forests, lakes, rivers and wildlife and to have comparison for living creatures".

These provisions are wide enough to empower the Government to do all that is necessary to do by legislative and administrative action to protect human environment. Yet much remains to be done particularly in the direction of implementation and identification of environmental problems.

PUBLIC AWARENESS

Public awareness about environment is at a stage of infancy. Development has paved the path for rise in levels of living but simultaneously led to serious environmental implications. Issues related to environment have often been termed as antidevelopment. To safeguard the human race it is required to maintain a balance between our needs and supplies so that the delicate ecological balance is not disrupted.

Today, environmental awareness needs to be created through formal and informal education to all sections of the society. Everyone needs to understand it because environment belongs to all. Various group activities and programmes can be use for raising environmental awareness in different groups of the society.

1. Among students through education- It is a encouraging step that now all over the country we are introducing environmental studies as a subject at all stages of schooling and college education following the directives of the supreme court.

2. Among the Masses : Media can play an important role to educate the masses on environmental issue through articles, rallies, plantation campaigns and success stories of conservation efforts. TV serials like Ract to save the planet, Terra-view, Bhumi and Programmes on National Geographic and Animal Planet have been effective in creating environmental awareness among the masses.

3. Among the plannes decision-makers : Since this group of society plays the most significant role in shaping the future of environmental resource management, it is very important to give them the necessary orientation through workshops and training programmes. Publication of ministry of environment & forest can also help in keeping this section abneast of new developments in the fields.

4. Role of Non Government Organizations : They can act as an effective an viable link between local population acid government. They can be very effective in organizing public movements for the protection of environment.

The chipko movement for conservation of trees by Dasholi Gram Swarajya Manda in Gopeshwar, Narmada Bachao Andolan by Kalpvriksh are some example where NGOs have played a landmark role for conservation of environment. The Bombay Natural History

Society (BHNS), a the World Wide fund for Nature-India and center for Science and Environment (CSE) are playing a pivtol role in creating environmental awareness. The recent report by CSE on more than permissible limits of pesticides in cola drinks sensitized the people all over the country.

CHAPTER-XV

Human Population and Environment

The *Homo sapience* has though achieved a dominating position in over the earth, but it still depends totally upon the various interactions among the different components of nature. His dependance on various bio-physical components of soil, water and air is inseparable and their relationships are represented in the forms of biotic community and ecosystems. However, in the process of development the various anthropogenic activities are causing environmental degradation which is leading to various social and environmental problems which need an immediate solution for the proper management and protection of environment.

The majority of countries which were in the process of development and improving the standard of living of their people and fulfilling their basic needs, considered environmental issues as luxury which only developed countries couln afford. Whereas the developed countries, which wereat the helm of environmental damage, were emphasizing for global action for the protection of environment. The situation has since been changed and developing countries have now realized the necessacity of environmental protection and many soicio-environmental issues are being taken care of. Some of these issues and problems are discussed here below.

POPULATION GROWTH: VARIATION AMONG NATIONS

During last century the human population on this earth has grown with a geometric rate. From the dawn of human history until the beginning of the last centuary the world population hovered around a few million, growing slowly than suffering setbacks due to epidemics. Human population changed from a condition of slow growth punctuated by setbacks to one of explosive geowth. The world population reached the 1 billion mark by the end of 1830. In in next 100 years it had increased to over 2 billion and in 1960, it reached 3 billion and by 1987 it crossed 5 billion. During the last couple of decades, the percentage rate of growth began to slow. But still with the large population base, even the lower rate of growth continue to add absolute numbers faster. It is estimated that by the end of 2010 the 8 billion mark will be crossed.

The population growth rates are different in different countries and on the economic scenario the world can be devided in to three categories:

1. Low income countries: It includes the India and other countries of East and Central Africa, some countries of South east Asia. These countries are also known as the third world.

2. Middle income countries: Many countries of Latin America, North and West Africa and East Asia.

3. High income countries: Thise are the developed nations and include United States, Canada, Japan, Australia and most of the Europian countries.

The highly develo--ed countries hold just 25% of the world population but they control about 80% of world 's wealth leaving remaining 20% for the developing countries which have 75% of world's population. During the last three decades development efforts resulted in some mitigation of this economc disparity but a considerable gap has been created between grownup and and poor countries. This trend is continuing currently. In the less developed countries, the population explosion is most intense and they are growing at rates which will double their populations in just 20-25 years. The case in the developed nation is just opposite and their population has approached almost a stability. The population data of some selected regions of the world are given in the following table:

Regions	Total Fertility Rate	Natality 1000-1 yr-1	Mortality 1000-1 yr-1	Gross National Product USD per capita
World	3.6	28	10	3,010
More developed	1.9	15	9	10,700
Less developed	4.1	31	10	640
Africa	6.3	44	15	620
Asia	4.3	32	11	1,480
Latin America	3.7	29	8	1,720
North America	1.8	15	9	17,170
USA	1.8	16	9	17,500
West Germany	1.4	10	11	12,080
France	1.8	14	10	10,740
England	1.8	13	10	8,920
Japan	1.7	11	6	12,850
Australia	1.9	15	7	11,910
Mexico	4	30	6	1,850
Brazil	3.4	28	8	1,810
Argentina	3.3	24	9	2,350
South Korea	2.1	19	6	2,370
China	2.4	21	7	300
India	4.3	33	13	270
Pakistan	6.6	43	15	350
Ethiopia	7.0	46	15	120

In most of the animals the environmental resistance keeps the natural populations in check and many environmental factors cause the dioff of a high percentage of the young before they reach reproductive age. It was the case with human beings also till the late 1800s. Epedemics odf diseases like smallpox, spotted fever, dysentery, diphtheria, measles and whooping cough took heavy tools of populations. After ajn epidemic, large families were left sometimes with no children at all. But after the innovation of vaccines for most of these diseases and improvement in sanitation situation has changed and now there is a dramatic change in the infant and child mortality. Crude death rates dropped from 40-50 per 1000 per year to as down as 10. Crude birth rate , however remained at 40-50. This dispariety inititated the population explosion . The fertility i.e. the number of children a couple has, is dependent on following two faxtors:

(i) The number of children a couple wants to have and

(ii) The availability of effective contraceptive techniques.

Effects on Environment

The growing populations of countries which do not have proper resources to meet their demands have a negative impact on the environment. For their day to day survival, they are overgrazing, overcultivating and over cutting forests for firewood, and commiting other acts which are directly affecting the ecological settings. This way, once a productive land is being deteriorated into worthless desearts which threatens not only the undeveloped or under-developed countries but the entire biosphere.

The production of food grains has increased per acre by use of irrigation, fertilizers, pesticides and by substituting highy yielding varieties. But these factors do have their own impacts on the environment. The irregation facilities on hand has inhanced the production but on the other it has also affected are third of world's irrigated land and farmers are forced to leave major tracts because of water logging and accumulation of salts in the soil.

The use of fertilizers has increaded many folds. In the beginning 15 to 20 additional tons were gained from use of fertilizers but situation has changed now and there is little to be gained by adding more. High level of fertilizers make plants more vulnerable to attack by pests, washing away of these fertilizers leads to water pollution. Similarly the use of pesticide have a lot of side effects to human and environmental health.

Control of Human population and family welfare programmes

The explosion in human population can be controlled substantially through suitable efforts in family planning, health care and education. With the crisis of overpopulation becoming more evident, people also have started realizing its ill effects and becoming more supportive to family planning programmes. Countries which have about 90% of world's population have family planning programmes and increasing attention is being given to such programmes. In 1987 a Statement on Population Stabilization was signed by 45 heads of developing countries. This statement reads in part: "Degradation of the world environment, income, inequality and the potential for conflict exist today because of over consumption and over population. If this over population growth continues, future generation of children will not have adequate food, housing, medical care, education, earth resources and employment opportunities....

We believe that the time has come to recognize the world wide necessacity to stop population growth within the near future and for each country to adopt necessary policies and programmes to do so..".

The family planning involves following:

(i) Educating people and society in general regarding the reproductive process and the pros and cons of various contrceptive techniques.

(ii) Providing contraceptives which individuals or couples may like to choose.

(iii) Counselling regarding achieving the best possible pre and post natal health of mother and child. Emphasis is given on good nutrition, sanitation and hygiene.

(iv) Counselling about advantages of spacing children. Emphassis is given on breastfeeding which provides both ideal nutrition and acts as a natural contraceptive since nursing mothers generally do not ovulate.

Health care is an integral part of family planning. In the developed countries it is given the utmost importance while in developing or in under developed countries the situation is very different. Here the poor health and mortality is very common among the mothers and children and involves poor nutrition and infectious deseases resulting from poor sanitation and hygiens. Once the couples begin to rear healthy and vigorous children the desire for more than two begins to drop. Similar is the case with education. The parents can see that the educated child will be more likely to achieve more success than an un-educated child. Once they start realizing this their desire for many children drops. In addition , education also allows women to persue other careers other than bearing children.

China, the largest developing country is the prime example of adopting proper family planning programmes. The leaders of the country realized that unless population growth is checked the country would be unable to live within its limits. Their goal was a one child family and to achieve this they used following incentives and deterents:

- Paid leaves to women who have fertility related operations like sterilization or abortion.
- A monthly subsidy to one child families.
- Job priority for single children
- Housing preference to single child families.
- Preferential medical care to parents whose only child is a girl.
- Payment of tax for a second child by the parents.
- Refund of bonus.
- Payment of higher prices for food for a second child.

POPULATION EXPLOSION: FAMILY WELFARE PROGRAMME

Population growth accelerated in middle of the 18th century with the advent of industrial revolution and associated improvements in agriculture. The growth of human population is a continuous process and in the last 10,000 yrs the population of the world has increased over a thousand fold, to about 4.5 billion. The current population of the world is estimated to be over 5.7 billions. According to United Nations estimates, the population growth rate at present hangs around 2.5% annually. In every 30 seconds, about 117 live babies are born and some 46 people die. This means over 75 million to be fed each year. Inspite of the implementation of population control programmes, in the next few decades there will be high growth rate, because of the effects of declining fertility. Another aspect of the population, which is considered to be effective in terms of its impact over the ecosystem is its spatial pattern both at macro and micro level.

The unequal distribution of population is the main cause of several environmental problems. On the one hand, there are countries like China and India, where population pressure is maximum and resources are limited, ad on other hand countries like USA, Canada where population is limited and resources relatively more are also present.

India is the best example of the relationship between population and environment. From ancient times much importance has been given to nature or environment. There was complete coordiantion and co-existence between man and nature but with rapid growth of population this balance has not only been disturbed but is now responsible for the degradation of environment.

India's population has grown very rapidly since the 1921, when medical treatment, sanitary improvement etc., started reducing mortality drastically and led to concomitant population explosion. The growth of population in India has created several environmental problems, some of which are:

- overall reduction in agriculture land
- uncontrolled urbanization.
- growth of slums
- problem of safe drinking water
- deforestation.
- problem of waste water and pollution.
- degrading quality of life.

To check on these problems there is urgent need to plan strategy to limit population. India is the first country, where family planning programmes initiated in 1950, under the directives of Government of India. The family welfare programme is recognized as a priority area in India and is a 100% centrally sponsored programme.

The main components of the programme include: to take all possible care for the safety of mothers and child birth, child health and, population control by adopting the family planning methods. In 1977 this programme renamed as family welfare programme. During 1951-61 a total of 2093 hospital providing facilities for fertility reduction programmes financial provisions are made under five year plans to boost family welfare programmes. Inspite of conducting campus etc. condoms, oral pills etc. are distributed free of cost in rural areas. Primary health centers in rural areas also providing facilities for male vasectomy and female tubectomy. With systematic efforts by government and communities including NGOs can make such programmes a grand success. Kerala has achieved stability in population growth. The stability was reached by achieving fairly good standards in health service, child care, women's education and land distribution through social engineering and government planning.

ENVIRONMENT AND HUMAN HEALTH

Environmental health basically deals with those forms of life substances, forces and conditions in the environment of human beings,which may exert an influence on man's health and well being. The earlier concept of health care reflected concern with communicable diseases and stressed for their prevention. But modern concept takes into account all aspects of human health including the attainment of desirable positive health values and prevention of undesired negative ones. The status of man's health represents the outcome of complex interactions between the internal biological system of a man his total external environmental conditions.

The studies of ecology and environmental science contributes to understanding of various aspects of human health including many diseases in several ways. The changes brought about by industrialization, urbanization, and population stress ultimately change the human environment. All over the world, more than 70% of diseases are water borne and water in turn is polluted by various activities of human beings. The major water borne diseases inIndia include cholera, dysentery and diarrhoea and these can be prevented by safe disposal of human excreata and adopting proper sanitary conditions. Similarly infaction of worms such as hookworms, round worms and guine worms also takes place through contaminated water. Skin diseases like seabies and eye infections are also included in this category.

There are a number of diseases caused by animals or transmitted by them to human beings. Many arthropods transmit diseses from animal to animal and from animals to man. Anopheles and Aedes aeqyptic carry the malarial parasites thus causing malaria in human beings. House flies are the most important vector of a number of human diseases like diarrhoea, dysentery, cholera, typhoid fever etc. Germs of the these diseases are carried by houseflies either by their mouth parts, legs or in the alimentary canal. Tse-tse fly carries Trypanosoma rhodesinsi and T. gambiense which cause sleeping sickness and if proper treatment is not given it may lead to death of the patient. Human louse suck the blood and transmits germs of certain diseases like relapsing fever, thyphus fever etc.

Rabbies commonly associated with dogs, has been a hazard. Contaminated soils offer the possibility of infection of many diseases. Walking barefoot on contaminated soil may provide an opportunity for entry of hookworms through the skin. Some people are more prone to communicable diseses which may be related to their general state of health, physical vitality, possible inherited factors and environmental conditions. This the reason that some persons become carriers but do not show symptoms of the diseses.

There are some other biological hazards in the environment besides the communicable diseases. There are scorpions, snakes and spiders whose venom is poisonous. Individuals become allergic to pollens of various plants, food materials like eggs etc., oil from plant origin, dust and other suspended material in the environment. These allergies may manifest themselves by skin eruptions swelling or other symptoms.

Similarly many chemicals released in to the environment pose a threat to human beings and their impacts on man depends upon the duration of exposure, concentration of chemicals and the susceptibility of an individual. Among the various toxic substances encountered through breathing are chemically active dust, gases, solvent vapour. Some are absorved through the skin directly. Similarly, chemicals concentrate in the various tissues of plants, animals and other organisms.

Any rise in the temperature of our environment changes the metabolic rates, oxygen consumption, heart beat and blood pressure.

An environment where mechanical noise and vibrations are prevalent may cause hearing loss and induce other psychological and physiological disturbances. Though, most of the time man gets adopted to his surronding conditions but this also causes stress resulting disorders of the body and mind.

India faces a disastrous double burden of diseases. Most of the old diseases continue to be rampant, while new ones are making rapid strides. Every third death in India is that of a child below the age of 5 who becomes a victim of combination of poverty, malnutrition, insanitary environment and polluted water drinking.

HUMAN RIGHTS

Human rights, belong to all people, or all people who are competent to exercise them. In contrast the right that only belongs to some people is termed a special right. The Human Rights may also be defined as- The rights of all individuals to certain fundamental freedoms enshrined in the Human Rights Act.

The rights people are entitled to simply because they are human beings, irrespective of their citizenship, nationality, race, ethnicity, language, sex, sexuality or abilities. Human rights become enforceable when they are codified as conventions, covenants, or treaties or as they become recognized as customary international law. Human rights are the universal rights which every human being should be entitled to enjoy and have protected.

Universal Human Rights and International Law: Largely through the ongoing work of United Nations, the universality of human rights has been clearly established and recognized in international law. Human rights are emphasized among the purposes of the United Nations as proclaimed in its charter, which states that human rights are "for all without distinction". Human rights are the natural born rights for every human being, universally. They are not privileges.

The charter further commits the United Nations and all member States to action promoting"universal respect for, and observance of, human rights and fundamental freedoms". As the corner stone of the International Bill of Rights, the Universal Declearation of Human Rights affirms consensus on a universal standard of Human rights. In the recent issue of a Global Agenda,

Charles Norchi points out that the Universal Decleration "represents a border consensus on human dignity than does any single culture of tradition".

Universal human rights are further established by the two international covenants on human rights (International Covenant on Economic, Social and Cultural Rights and International Covenant on Civil and Political Rights), and the other international standard-setting instruments which address numerous concerns, including genocide, slavery, torture, racial discrimination against women, rights of child, minorities and religious tolerance.

These achievements in human rights standard setting span nearly five decades of work by the United Nations General Assembly and other parts of the UN system. As an assembly of nearly every state in the international community, the General Assembly is a unique representative body authorized to address and advance the protection and promotion of human rights. As such it serves as an excellent indicator of international consensus on human rights.

This consensus is embodied in the language of United Decleration itself. The universal nature of human rights is literally written into the title of the Universal Decleration of Human Rights. Its preamble proclaims the Decleration as a "common standard achievement for all people and all nations". The statement is echoed most recently in the Viena Development and Programme of Action, which repeats the same language to reaffirm the status of Universal Decleration as a "common standard" for every one. Adopted in June 1993 by the UNWorld Conference on Human Rights in Austria, the Vienna decleration continues to reinforce the universality of human rights, stating," All human rights are universal, indivisible and interdependent and interrelated". This means that political, civil and cultural, economic and social human rights are to be seen in their entirety. One can not pick and choose which right to promote and protect. They are all of equal value and apply to every one.

As if to settle the matter once for all,the Vienna decleration states in its first paragraph that "the universal nature of , all human rights and fundamental freedoms is beyond question. The unquestionable universality of human rights is presented in the context of the obligation of States to promote and protect human rights.

Human rights are the birth rights of every person. If a state dismisses universal human rights on the basis of cultural relativism, then rights would be denied to the persons living under that State's authority. The denial or abuse of human rights is wrong, regardless the culter of violeter.

Human Rights, Cultural Integrity and Diversity

Universal human rights do not impose one cultural standard, rather one legal standard of minimum protection necessary for all human dignity. As legal standard adopted through the United Nations, universal human rights represent hard won consensus of the international community, not the cultural imperialism of any particular region or set of traditions. Like most areas of international law, universal human rights are a modern achievement, new to all cultures. Human rights are neither representative of, nor oriented towards, one culture to the exclusion of others. Universal human rights reflect the dynamic, coordinated efforts of the international community to achieve and advance a common standard and international system of law to protect human dignity.

Inherent Flexibility

Out of this process, universal human rights emerge with sufficient flexibility to respect and protect cultural diversity and integrity. The flexibility of human rights to be relevant to diverse cultures is facilitated by the establishment of minimum standards and incorporation of cultural rights. The instruments establish minimum standards for economic, social, cultural and political rights. Within this frame work, States have maximum room for cultural variation without diluting or compromising the minimum standard of human rights established by law. These minimum standards are infact quite high, requiring from the State a very high level of performance in the field of human rights.

Most directly, human rights facilitate respect for protection of cultural diversity and integrity, through the establishment of cultural rights embodied in instruments of human rights law. These include: the International Bill of Rights; the Convention on Rights of Child; the International Convention on the Elimination of All Forms of Racial Descrimination; the Decleration on Race and Racial Prejudice; the Decleration on the Elimination of All Forms of Intolerance and of Discrimination Based on Religion or Belief; the

Decleration on the Principle of International Cultural Cooperation; the Decleration on the Rights of Persons Belonging to National or Ethenic, Religious and Linguistic Minorities; the Decleration on the Right to Development; the International Convention on the Protection of the Rights of All Migrant Workers and Members of their Families; and the ILO Convention No. 169 on Rights of Indegenous and Tribal Peoples.

Human rights which relate to cultural diversity and integrity encompass a wide range of protections, including: the right to cultural participation, the right to enjoy arts, conservation, development and diffusion of culture; protection of cultural heritage; freedom for creative activity; protection of persons belonging to ethenic, religious, or linguistic minorities; freedom of assembly and association; the right to education; freedom of thought' conscience or religion; freedom of opinion and expression; and the principle of non-discrimination.

VALUE EDUCATION

The issue of value education has been highlighted as one of the priorities in the National Policy of Education (!986), which declares "The growing concern over the erosion of essential values and an increasing cynicism in society has brought to the focus the need for readjustment in the curriculum in order to make education a forceful tool for the cultivation of social and moral values". Thus the value education refers to a wide gamut of learning and activities ranging from training in physical health, mental hygiene, etiquettes and manners, appropriate social behaviours, civic rights and duties

Environmental education is a process of recognizing the value and various conceptions with the aim of determining the skills and approaches for understanding the connection between man, his culture and the biophysical environment. This certainly involves the adoption of decisions and the elaboration of a code of behavior relating to the quality of the environment. In 1975 an international workshop was organized by UNESCO on Environmental Education-The Belgrade Charter at Belgrade in Yugoslavia. It formulated the guiding principles to achieve the Objectives of Stockholm conference in 1972. The Belgrade workshop was followed by an international conference on Environmental Education held in the year 1977 at Tbilisi in The then USSR to follow the footsteps of Belgrade workshop.

The most relevant of the subject of Environment education is a Chinese perception about education which says "If you plan for a year, plant rice, if you plan for 10 years, plant trees, but if you plan for one hundred years, educate the people."

Environmental education includes teaching on determining certain values and the ability of clear thinking of complicated environmental problems and the best insurance for the environment is a commitment on behalf of the public to prevent the deterioration of air, water and land. It is a way to achieve targets fixed to protect the environment and should constitute a comprehensive life long education.

In this age of rapid industrialization many people recognize the urgent need for environment education. The chief goal of environmental education is to develop our population which is aware of and is concerned about the environment and its associated problems, and who has the knowledge, skill and attitude with commitment to work individually and collectively towards the solutions of current problems and prevention of new ones. The goals of environment education according to Tbilisi Declaration are: To foster a clear awareness of, and concern about, economic, social, political and ecological interdependence in urban and rural areas; to provide every person with opportunities to acquire the knowledge, values, attitude, commitment and skills needed to protect and improve the environment and; to create new patterns of behavior of individuals, groups and society as a whole towards the environment.

A number of objectives have been assigned to environmental education to help social groups and individuals towards:

(i) Awareness: To acquire awareness and sensitivity to the total environment and associated problems.

(ii) Knowledge: To help individuals and social groups to acquire basic understanding of total environment .

(iii) Attitude: To help individuals and social groups acquire social values, strong feelings of concern for the environment and the motivation for actively participating in its protection and improvement.

(iv) Skills: to help individuals and social groups acquire the skills for solving environmental problems.

(v) Evaluation ability: To help individuals and social groups evaluate environmental measures and education programmes in terms of ecological, social, political and education factors.

(vi) Participation: To help individuals and social groups to develop a sense of responsibility and urgency regarding environmental problems to ensure appropriate action to solve those problems.

A three tire classification of environmental education programme has been proposed by Newman in 1981:

(a) Environmental studies: Concerned with the environmental disturbance and minimization of their impacts through changes in the society.

(b) Environmental science: Deals with the study of the processes in water, air, soil and organisms which lead to pollution or environmental damage and to know a scientific basis for establishing a standard which can be considered acceptability clean, safe and healthy for human and natural ecosystem.

(c) Environmental engineering: to study the technical processes which are used to minimize the pollution and the assessment of impact of these on environment.

Indian Scenario

The education commission (1964-66) correctly observed that: The school curriculum is in a state of flux all over the world today. In developing countries, it is generally criticized as being in adequate and outmoded and not properly designed to meet the needs of modern times— a new reform movement has started which may bring in sweeping curricular changes in school educatiobn— the tremendous explosion of knowledge in recent years and reformation of basic concept in physical, biological and social sciences have brought in to sharp relief of adequacies of existing school programmes.

As per the recommendations of the Education commission: The aim of teaching science in primary school should be to develop proper understanding of main facts, concepts, principles and processes in the physical and biological environment.

At the higher primary stage environmental activities will lead to the study of natural and physical sciences, history, geography and civics— and the practices of healthy living.

At the secondary stage the needs of development of certain skills, attitudes and qualities...

The National Policy on Education, 1986 says that there is a paramount need to creat4e a consciousness of environment. It must permeate all ages and all sections of society beginning with the child. Environmental consciousness should be a prt of teaching in schools and colleages. This aspect will be integrated in the entire educational process.

Today the environmental education is being imparted at each level and special caps and workshops are being organized ~~to spread~~ the message. The subject is being thought right from the primary to the University level with many special institution setup exclusively for this purpose.

Environmental awareness

Effective implementation of environmental management and conservation programmes depends on awareness raising and training in the relevant areas. Without an understanding of how to conserve our resources, or indeed, why they must do so, only a few communities would be motivatede to participate in conservation and management programmes. Sensitizing people by organizing workshops, symposia and through various educational programmes at different levels will help in spreading this awareness. The task of education, training and information is to clarify the environmental cosequences of decisions and the right solution modes for everyone. The mass media plays a crucial role in forming the attitude of the public to the environment. At present though the press, radio and television cover environmental issues, the impact seems to be insignificant as they are unable to counterbalance the deluge of opposite information from the same source. Improvement of the situation is a strategic issue but it means not only the increase of frequency but quality also. Some of the daily news papers have statrted columns on environment and environmental related problems. They give regular information on incidents and opinions. Now various programmes highlighting different aspects of environment viz biodiversity, pollution and scarcity of resources are being regularly shown on the various television channels.

Education begins at nursery schools and initiatives like competition for the creation of original stories and poems to establish and improve the environmental approach and emotions of nursery children may be taken at this stage. For secondary and higher level there are a number of activities related with nature conservation and environment protection where the students can be involved and trained as future environmentalists.

In India, the Ministry of Environment and Forests statrted a National Environmental Awareness Campaign (NEAC) in 1986 for creating environmental awareness at all levels of society. Many organizations located in different parts of the country have been designated as regional resource agencies(RRAs) for assisting the Ministry in conducting this campaign. To impart environmental education and to encourage and mobilize participation of school children in various environmental conservation activities in their localities, the Ministry has been providing financial assistance for setting up of Eco-Clubs in schools. The following themes have also been identified for the mass awareness campaign: Clean air; clean water; biodiversity; conservation and management of waste.

HIV AND AIDS

AIDS is the acquired immunodeficiency syndrom and is caused by HIV or Human Immuno deficiency virus. This virus impairs the immune system, making a person prone to infections and diseases. HIV is transmitted through exposure to the body of a person infected with HIV. This exposure most commonly occurs during unprotected sex, by sharing needles, through blood transfusions, or by contact with open wounds. Babies born to women with HIV can also become infected. A gradual deterioration of immune function takes place. Most notably, crucial immune cells called CD4+T cells are disabeled and killed during the typical course of infection. These cells sometimes called T- helper cells, play a central role in the immune response, signaling other cells in the immune system to perform their special function. A healthy, uninfected person usually has 800 to 1,200 CD4+T cells per cubic millimeter of blood. During HIV infection, the number of these cells in a person's blood progressively declines. When a person's CD4+T cell count falls below 200 /mm3 ,he or she becomes particularly vulnerable to the opportunistic infections and cancers that typify AIDS, the end stage of HIV disease. People with AIDS often suffer infections of lungs, intestinal tract, brain, eyes and other organs as well as debilitating weight loss, diarrhea, neurologic conditions and cancers such as Kaposi's sarcoma and certain types of lymphomas.

Most scientists think that HIV causes AIDS by directly inducing the death of CD4+T cells or interfering with normal function, and by triggering other events that weaken a person's immune function.

HIV belongs to a class of viruses called retroviruses. Retroviruses are ribonucleic acid (RNA) viruses, and in order to replicate they must make a deoxyribonucleic acid (DNA) copy of their RNA. It is the DNA genes that allow the virus to replicate. Like all viruses, HIV can replicate only inside cells, commandeering the cells machinery to reproduce. However, only HIV and other retroviruses, once inside the cell, use an enzyme called reverse transcriptase to convert their RNA into DNA, which can be incorporated into the host cell's genes.

HIV belongs to a subgroup of retroviruses known as lentiviruses or slow viruses. The course of infection with these viruses is characterized by a long interval between initial infection and the onset of serious symptoms.

HIV/AIDS Worldwide: An estimated 37.8 million people worldwide- 35.7 million adults and 2.1 million children below the age of 15 years were living with HIV/AIDS by the end of 2003. Approximately two third of these people live in Sub-Saharan Africa another 20% live in Asia and the Pacefic. Worldwide, approximately 11 of every 1000 adults aged 15 to 49 are HIV infected in Sub Saharan Africa, about 7.5% of all adults in this group are HIV infected . Woman account for nearly half of all the people world wide living with HIV/AIDS. An estimated 4.8 million new HIV infections occurred worldwide during 2003, that is about 14,000 infections each day. More than 95% of these infections occurred in developing countries. More than 20 million people with HIV/AIDS have died since the first AIDS cases were identified in 1981. In 2003 alone, HIV/AIDS associated illness caused the deaths of approximately 2.9 million people worldwide, including an estimated 490,000 children below the age of 15 years.

WOMEN AND CHILD WELFARE

The role of women in environmental issues was highlighted in the UNEP State of the Environment Report 1988, and in the the book Women and Environment in the Third World by Joan and Davidson in 1988. The topic has been a focus of discussion on various plateforms world over. The increasing deterioration of environment is having a great impact on all its constituents and for women worldwide, it has a particular significance. In their book, Women in the World, S. Jovelan et al have stated that "for women there are no developed countries". It very clearly implicates that the women everywhere work longer hours than men, are less well nourished and in poor health. Because of their routine tasks of caring their family and community, women in most of the developing and underdeveloped countries get affected by their environment and vice versa. In such countries natural resources are collected by woman folks. Woman carry multiple roles in the society. They may produce food and income. Because of their special responsibilities in the family and community, women have the potential to influence community attitudes and activities towards the environment.

The knowledge of various environmental factors acquired by many women is both useful and detailed. They are not only the resource managers but also the victims of environmental mismanagement. Their local environmental interests have been affected by inappropriate development. Projects and programmes have also failed in addressing the specific needs of women, although their role in productive sectors such as agriculture, water, forestry, energy and human settlements is well documented. The great movements like Chipko in Uttaranchal and save Narmada in Madhya Pradesh were initiated by the women and probably no other group of our society is affected so much so by environmental degradation than the poor women of our villages.

The UN General Assembly in 1959 adopted The Declaration of the Rights of Child. After the UN Convention, it became an

International Law in the year 1990, which promotes and protects the well being of child in the society.

ROLE OF INFORMATION TECHNOLOGY IN ENVIRONMENT AND HUMAN HEALTH

The modern era is now known as the era of Information and Technology where the distances between different parts of the world have been reduced and all the knowledge is available to every one sitting at any corner of the world. The scope of Information and Technology in the field of education , may be environmental education or the medical education is tremendous. Now the information on different areas of environment, medical and health is available at a click of button. Development of internet facilities, world wide web, remote sensing and geographical information system and colleting the information through the use of satellites have made it possible to generate a large amount of information within no time.

Development of a large number of soft ware has added another dimension to this facility. Now the operations and medical treatments are being performed even by seating in a different country. People are taking the benefits of telemedicines. Similarly by the use of remote sensing data and geographical information system data are being gathered on different environmental aspects like, forest fire, forest degradation, floods, draughts, desertification, urban sprawl, new mineral resources, water resources, energy reserves and on many other geological aspects. Now a days the BPO i.e Business Process Outsourcing has become an important source of transer of technology in respect of environment and health. BPO can be defined as the transefer of an organization,s non core but critical business functions to an external vendor who uses an information technology based service delivery. Since delivery is IT base, BPO is also referred as IT Enabled Service (ITES). All through the world this facility is helping in the fields of pharmaceuticals, health, environmental protection, insurance, telecome and banckining etc. by decimination the data and technologies from one corner to another corner of the world.

CHAPTER-XVI

FIELD WORK

VISIT TO A LOCAL AREA TO DOCUMENT ENVIRONMENTAL ASSETS

To better understand the ecology, environment and role of various factors on the environment, a visit of the students can be planed to the nearby areas like a forest, a river, a zoo, botanical garden, industries emitting gaseous pollutants and areas which have been encroached by the people after disturbing the natural habitats and resources. A large number of natural resources are present in each of the ecosystem. A field visit may give an insight to the students about the area, the type of ecosystem, the various components of the ecosystem and different resources that are available in the ecosystem. Similarly by making a field visit students can better understand the concept of a community, which is an important ecological principle and emphasizes the orderliness that exists in the assemblage of a variety of organisms living in any habitat. Most communities exhibit a pattern or structure in the arrangement of the components. The structure of a community may be in the form of vertical stratification (as in forest community), horizontal zonation (as in marine intertidal community), or in functional pattern relating to activity, food web, or social behavior of organisms.

Students may prepare field reports and describe the facts they have seen and the impacts of different ongoing activities, which they feel will be on the ecology and the environment of their area.

VISIT TO A RIVER SIDE

Depending on the situation, students may visit to any river or pond of their area and may note down their observations as under:

(i) Name of the river, place of its origin, its route and also cities and industries based on its bank.

(ii) Observe the different attributes of river water like clarity of water, any vegetation that is present in the river bed, light penetration, transparency of water body using a Sacchi disc.

(iii) The temperature of the water may be observed by using a thermometer.

(iv) The productivity of the river can also be measured by various methods e.g., chlorophyll method, light and dark bottle methods etc.

(v) Mixing of any other stream or effluents in the main river must also be noted.

(vi) If effluents are coming to the river, their source and and the visual differences due to such mixing may also be noted.

(vii) The diversity of fauna like fish, insect etc. in the river may also be noted

(viii) Use of water for different purposes like agriculture, industry and drinking purposes may also be noted.

(ix) If possible, some of the chemical parameters may also be recorded to determine the purity of water like, pH, alkalinity and dissolved oxygen.

(x) Inference on impact of human activities on water body can also be drawn after recording the history of the river.

Exercise

Study the biotic components of a river or pond ecosystem:

Requirements: Collection nets, glass jars, hand lens, metallic chains with hook.

Procedure: The plant and animal species growing near the margins of the river or pond can be easily picked up by hand and placed in polythene bags. The large sized plants , which are submerged can be collected using hooks fitted with metallic chains. Students can collect algae, protozoan etc. from the different depths of the river or a pond with the help of plankton nets and store in bottles. All the collections are brought to the laboratory and identified with the help of available keys. All the species so identified can be placed into following heads:

(i) Producers: The plants which bear green colour are put in this category. These may be:

(a) Submerged species like *Hydrilla, Vallisneria, Ceratophyllum* and *Chara* etc.

(b) Freefloating species like *Spirodella, Pistia, Wolffia, Eichhornia, Ezolla* and *Lemna.*

(c) Rooted but floating plants may include *Nelumbo, Nymphaea* etc.

(d) Amphibious are like *Typha, Ranunculus* etc.

(e) Phytoplanktons are the members of chlorophyceae, maxophyceae, bacillariophyceae and xanthophyceae.

(ii) Consumers: These may be : Primary consumers like mollusca, insects and zooplanktons; secondary consumers like small fish, some insects and frogs etc. and tertiary consumers may include big fish and birds.

(iii) The Decomposers: These may be different types of bacteria, fungi etc.

VISIT TO A FOREST

(i) Forest type, dominant species, their density, types of shrubs and herbs may be observed and noted

(ii) Is here any pressure on the forest resources like collection of wood, fodder, grasses, minor produces etc. may be recorded.

(iii) A list of wild animals which have been reported from the forest may be prepared and accordingly the animals seen during the visit may be marked.

(iv) Fragmentation in the forest, any other illegal activity in the area as observed may be recorded.

Exercise

To study the vegetation and determining the frequency of individual species in a forest area.

To make such studies, we normally follow the line transect method. A measuring tape is run across the vegetation; all the species touching the tape in alternate segments of a predetermined length are noted. Depending upon the area to be covered, any number pf transects may be laid down at random across the forest area at different sites. If one meter long segment is the unit of study and three plants of any one species are touching the segment, we say that the numerical strength of that species is three individuals per meter.

Like, wise observations may be made and noted down and frequency and density of different species of plants may be determined by using the following formula:

Frequency (%) = 100 X total no. of quadrates (sampled segments) in which a particular species recorded / total number of segments studied.

Density = Total number of individuals of a species/ total number of segments studied

VISIT TO AN INDUSTRIAL SITE TO STUDY THE POLLUTION

(c) Name of the industry, year of its establishment, products of the industry, manufacturing capacity and by-products from the industry.

(ii) Different pollution emitting aspects of the industry may be observed. First see if there is a stalk at the top of the building and emitting a large amount of smoke or toxic gases. Ask for the types of gases and see if these have a foul odor. Ask if the industrialist has put some scrubbers or other controlling devices to control or check the relies of these gases in the atmosphere.

(iii) Look for the effluents that are being generated by the industry, its colour, smell, and the site where it is being disposed off. If possible determine the pH, temperature and dissolved oxygen of the effluents and compare with the river or pond water.

(iv) Find out the working conditions inside the factory. See what is the level of noise, ambient temperature and dust inside the working premises. Ask some simple questions to the workers like how many hours do they work, if they have any health problems etc.

(v) Look for a green belt around the industry and find out the colour of leaves, branches and fruits of the trees.

TO VISIT AN AGRICULTURAL FIELD AND STUDY THE TEMPERATURE, COLOUR AND SOIL TEXTURE

Temperature of soil is measure by a special type of thermometer known as soil thermometer, which has vertical arm with bulb at one end and a dial with deflection needle on the other end. The bulb of the thermometer is buried at different depths i.e., 2 to 15 inches of the soil. Note down the temperature on the dial andvariations at different depths may be recorded.

To determine the colour of the soil we need a Munsell's soil colour chart. The soil is spread uniformly over a card board sheet. The soil partcles are now matched with chips of a number of colours in the Munsell's soil colour chart and the colour of the soil is determined accordingly.

To determine the texture of the soil, sieves of different mesh size are used. This is a standard method. The soils passed successively through a series of sieves with different mesh size and different fractions of soil are separated from the sample, find out relative proportion of each fraction and determine the texture of soil according to this proportion.

Diameter (mm) of soil particle	Name given to the particles
Below 0.002	Clay
0.002- 0.02	Silt
0.02-0.20	Fine sand
0.20-2.00	Coarse sand

STUDY OF SOME SIMPLE EQUIPMENT

1. Jackson's Turbidity Meter

It is used to measure the turbidity of water. It has three parts; stand, candle and calibrated glass tube. The turbidity of the water is determined by filling the glass tube with given sample of water slowly till the light of candle disappears. The reading in cm given on the tube is noted and is converted in to JTU with the help of conversion table.

2. Secchi Disc

It is a circular disc with alternate black and white column with a rope attached at the centre of disc and used to measure the transparency of water.The diameter of the disc is 20 cm The disc is slowly lowered in the water till it just disappears from view, this depth is recorded. Now lower it just a little further and then gradually raise it until it just reappears, this depth is also recorded. The transparency of water is calculated by using the formula:

Transparency = a+b/2; where a= length of the rope at which disc disappears and, b= length of rope at which disc re-appears.

3. Plankton collection net

It is a conical structure with a steel ring at the top and a silk cloth of different mesh size fixed on the ring. At the bottom of the net a collection tube of known volume is attached. Measured amount of water is collected through net during the collection and the sample so collected is preserved in formalin for further studies in the lab to determine the composition of planktons.

STUDY OF SOME PLANTS, INSECTS AND BIRDS

We all know that a variety of plants and animals exist around us, This variability among the living organisms is known as bio-diversity. A large number of plants and animals provide us materials which are put to different uses like food, medicines, energy, timber and a lot of other uses. It will be very useful if we know about some of the plants, insects and birds of the area where we live:

Plants

Plants are not only the primary producers but also provide other materials to us which are of great economical value. The trees like Sal, Shisham, teak, deodar provide us timber; many varieties like Mellitus provide us fodder for our animals. Some plants like jatamansi (*Nardostachys grandiflora*), Daruhaldi (*Berberis asiatica*), Sadabahar etc. provide us medicines for different ailments.

Insects

Insects are both useful and harmful to us. We may find a large number of insects around us. House fly, mosquito, termites, bed bug, head louse, cockroach, and grasshoppers etc. are the insects which cause damage and spread many diseases but Honey bee, silk worm, butterflies and lack insects are the one which not only give us products which are used in a variety of way but also these insects serve the ecosystem by pollinating a large variety of flowering plants.

Birds

Birds are other important creatures which we find very fascinating. A large number of birds visit to our area every day and serve the environment in different manners. The crow, house sparrow, and munias the most common examples which we see every day. The vulchers and parrots are the one which are seen on special occasion. We may prepare a list of different bird species which visit to our area and find out that how many of them are useful to us and the manner by which some of them cause damage to us.

QUESTIONS FOR REVISION

1. Explain the statement that ecology is a multidisciplinary subject.
2. What are the various approaches to study the ecology and environmental science.
3. Write in details about the scope and importance of environmental science.
4. Describe how public awareness can help to protect the environment.
5. What is a natural resource. Describe the renewable and non renewable resources.
6. What are the two major methods of conserving natural resources.
7. What role an individual can play in conserving the natural resources.
8. Describe the various uses of forest. How forest resources are being over exploited.
9. What are the important causes of deforestation
10. Write short notes on following:
    i. Shifting cultivation
    ii. Effects of deforestation on tribal people
    iii. Social forestry
    iv. Over exploitation of surface and ground water
    v. Conflicts over water
11. Explain with examples, how big dams have degraded our forests and land
12. What are the various environmental effects of extracting and using mineral resources.

13. Discuss briefly the various non renewable resources
14. Discuss how overgrazing and agriculture are contributing to environmental degradation
15. What are the different structural components of an ecosystem
16. Discuss the laws governing the energy transformation in an ecosystem
17. What do you mean by food chain and food web.
18. Describe the various biotic and abiotic components of an ecosystem you have studied
19. discuss the process of ecological succession
20. Explain the following:
    i. Species diversity, genetic diversity and ecosystem diversity
    ii. Value of biodiversity
    iii. India as a mega biodiversity nation
    iv. Hot spots of diversity
21. Describe the various threats to biodiversity
22. Give a brief account of endemic fauna of India
23. Give a list of the most widely spread and seriou pollutants
24. What are the important sources of air pollution
25. Enumerate the various effects of air pollution.
26. What methods can be employed to control the air pollution.
27. Explain how water gets polluted
28. Define the term thermal pollution. What are the various effects of thermal pollution

29. Describe the sources of marine pollution

30. What is a solid waste. Describe the various sources of solid waste

31. Describe the various effects of solid waste and its management practices

32. Write notes on Bhopal gas tragedy and Chernobyl nuclear disaster

33. Describe the various effects of soil pollution

34. Discuss how noise pollution affects human beings

35. What role an individual can play in preventing the environmental pollution

36. What is a disaster. How can these be managed

37. Give a brief account of Environment (Protection) act, 1986.

38. Describe the various issues involved in enforcement of environmental legislation

39. How variation of population among nations is responsible for environmental problems.

40. Describe the role of environment on human health

41. Give a brief description of HIV/AIDS

42. Discuss the issue of value education

43. What are the objectives which can be assigned to environmental education

44. Describe the role of women in environmental conservation

45. What role information technology can play in protection of environment and human health

GLOSSARY

Abiotic- Which are non living and independent from living things.

Adaptations: A change in structure or function that produces better adjustment of an organism to its environment, thus enhancing its ability to survive and reproduce.

Aerobic organism :Organisms which need oxygen to survive.

Age- The age of an individual is the amount of time that has passed since its birth.

Air- The mixture of gases viz. 78% nitrogen, 21% oxygen and 0.03% carbon dioxide, making up the atmosphere.

Annual- Any plant or organism that lives for only one season.

Atmosphere- It is the gaseous mantle which envelops the hydrosphere and lithosphere.

Benthic plants_ These are aquatic plants that grow attached to or rooted in the bottom.

Bioaccumulation- The accumulation of higher and higher concentration of a potentially toxic chemical in living beings.

Bioconversion-Use of biomass as fuel.

Biodegradable- Material which can be consumed and broken down to natural substances.

Biodiversity- Variety and variability of life on the earth.

Biological Oxygen Demand (BOD)- This is the amount of dissolvrd oxygen consumed by biological organisms in the process of decomposing organic material present. Any water body with higher BOD has a poor water quality.

Biomass: **Dry** weight of all organic matter in plants and animals in an ecosystem.

Biome- It is a group of eco-systems that are related by having a similar type of vegetation and with similar climatic conditions.

Bio-sphere- It is that part of earth where life can exist and is the sum total of all the biomes and smaller ecosystems.

Biotic potential : Maximum rate at which the population of a given species can increase when there are no limits of any sort on its rate of growth.

Birth Rate- The change in the number of births in a population measured over a short time interval.

Carrying capacity- The maximum number of individuals that can be supported in a population that is growing according to the logistic growth equation. This limit reflects the availability of food, space and other resources in the environment.

Community- It is the local association of several populations of different species living in a prescribed area or habitat.

Conservation- It is the management of resources in such a way so that it will continue to provide maximum benefits to humans for a long term.

Consumers- Organisms that derive their energy from feeding on other organisms or their products.

Darlington's rule- It states that on oceanic islands, a tenfold increase in island area is needed for each doubling of number of species.

Death Rate-It is the change in the number of deaths in a population measured over a short time interval. In the Latka-Volterra predation model, the death rate is the instantaneous death rate for the predator population in the absence of the victim population.

Decomposers- Which convert complex substances into simpler components.

Disasters: Disruption to normal pattern of life, which is severe, sudden, unexpected and widespread.

Dissolved Oxygen- Amount of oxygen dissolved in water.

Draught: Unusual dry period which results in a shortage of water in the rivers, reservoirs and wells.

Earthquake: It is a set if vibrations produced in the earth's crust when rocks in which elastic strain has been building up, suddenly ruptures and then rebound.

Ecology- Study of the relations of organisms to their environment.

Ecological niche – It is an animal's place in the biotic environment with relation to its food and enemies.

Ecosystem- A grouping of plants, animals and other organisms interacting with each other and their environment.

Ecotone- It is the transitional zone or junction zone between two or more diverse communities.

Energy- Ability to do work.

Enthalpy- It is a change in the amount of energy in the form of heat liberated or absorbed by the system during physical or chemical changes.

Entropy- Measure of energy not available due to transformation of energy at each successive trophic level in the food chain.

Environment- The combination of all things and factors external to the individual or a population.

Environmental Impact- Effects on the natural environment caused by human actions.

Environmental Impact Assessment- A study of the probable environmental impacts of a developmental project on environment.

Emigration- Individuals leaving a population and traveling to another location. It decreases the size of population.

Environmental stochasticity- Uncertainty due to variation in environmental conditions.

Euphotic zone- The depth from the surface to the limit of adequate light to support photosynthesis.

Eutrophication- It refers to nutrient enrichment that promotes the growth of phtoplanktons in a water body. It results in the turbidity, death of benthic plants and depletion of dissolved oxygen.

Exponential population growth- A simple model of population growth in which the population growth rate is the product of the current population size and the instantaneous rate of increase. Although, no population in nature ever shows exponential growth for very long, all populations have the potential to increase exponentially because each individual can leave more than one offspring in the next generation.

Extinction-The disappearance of all the individuals of a species.

First law of thermodynamics- Energy is neither created nor destroyed but is converted from one form to another.

Flood: A high flow of water which overtops the natural or artificial banks of a river.

Food chain- The transfer of energy and material through a series of organisms.

Food web- It is the combination of all the feeding relationships that exist in an ecosystem.

Fossil fuel- Energy rich materials derived from prehistoric photosynthetic production of organic matter on earth like natural gas, petroleum etc.

Green house effect- An increase in the atmospheric temperature because of increasing amount of carbon-dioxide and certain other gases which absorb and trap the heat normally radiated from the earth's surface.

Habitat- The specific environment in which an organism lives.

Humidity- The amount of water vapour in the air.

Hydrosphere- The sum total of all liquid components that is the water in oceans, lakes, rivers and on lands.

Immigration- Individuals entering a population from another location. It increases the size of population.

Lithosphere- It comprises the solid components i.e., the rocky substances of the continents.

Metapopulation: A group of several local populations that are linked by immigration and emmigaration.

Noise- Unpleasant or unwanted sound that causes discomfort.

Non-Renewable- resources that exist in finite quantity in the earth's crust and are not replenished by natural processes as they are mined.

Oligotrophic- The natural condition of lakes and estuaries where water is nutrient poor.

Pollution- It is the unfavorable alteration of our surroundings or contamination of air, water or soil with undesirable amounts of material.

Pollutant- A substance the presence of which causes pollution.

Population- It is a group of individuals, all of the same species, that live in the same place.

Population growth rate- It is the change in population size during a small interval of time.

Oil field: The area in which exploitable oil is found.

Oil Shale: A natural sedimentary rock that contains a material which can be extracted and refined in to oil and oil products.

Oligotrophic: A lake the water of which is nutrient poor and does not support the phytoplankton but supports only the submerged aquatic plants.

Organism: Any living thing-plant, animal or microbe.

Remote sensing: It is the science and art of obtaining information about an object, area or phenomenon through the analysis of data acquired by a device that is not in contact with the object, area or phenomenon under investigation.

Renewable- Biological resources that may be renewed by reproduction and re-growth.

Second law of Thermodynamics- It states that in every conversion, some heat always escapes from the system or in other words in every energy conversion, a portion of energy is lost.

Soil profile- A description of the different, naturally formed layers within a soil.

Succession- It is the orderly process of community changes, which are directional and, therefore predictable.

Volcano: Opening in the surface of the earth from which molten rocks, magma and gases escape.

Wildlife- All animals and plants that are not domesticated.

Suggested Readings

Arun Kumar (1999). Environmental problems protection and control.Anmol publications Pvt, Ltd. New Delhi.

Chaturvedi, R. G. and Chaturvedi, M. M. (1994). Law on Protection of Environment and Prevention of Pollution. The Law Book Company (P) Ltd. Allahabad.

Clarke, G. L. (1954). Elements of Ecology. John Wiley & Sons, Inc. New York.

Gotelli, Nicholas J. (1998). A primer of Ecology. Sinauer Associates, Inc. Sunderland, Massachusetts.

IUCN, 2000. IUCN red list of threatened species. Compiled by C. Hilton-Taylor. IUCN/SSC, Gland, Switzerland.

John, R. Jenson(2003). Remote Sensing of Environment-A Earth Resource Perspective. Printice Hall Series.

John, Glasson, Rikki Therivel and Andrew Chadwik (2000). Introduction to Environmental Impact Assessment. T.J. International Ltd., Padstow.

Joshi, P. C. and Joshi, Namita (2004). Biodiversity and Conservation. APH Publishing Corporation, New Delhi.

Joshi, P. C. and Joshi, Namita (2005). A text book of ecology and Environment. Himalaya Publishing House, New Delhi

Kendeigh, S. C. (1974). Ecology with special reference to animal and man. Prentice Hall, New Jersey

Larry, W. Canter (2000). Environmental Impact Assessment. McGraw-Hills, New York.

Lillesand, Thomas M and Kiefer, Ralph W. (1994). Remote sensing and image interpretation. John Wiley and Sons, Inc. New York.

Misra, R. (1974). Ecology Work Book, Oxford and IBH New Delhi

Nebel, Bernard J. (1990). Environmental Science- The way the world works. Prentice Hall, New Jersey.

Odum, E. P. (1963). Ecology. Holt Rinehart and Winston, New York

Odum, E. P. (1966). Ecology. Modern Biology Series, Holt, Rinehart and Winston, New York.

Odum, E. P. (1971). Fundamentals of Ecology. Saunders College Publishing, Philadelphia.

Patra, A. K. (1997). Environmental studies (Man and Environment).Kalyani Publishers, New Delhi.

Pullin, Andrew S. (2002). Conservation Biology. Cambridge University Press.

Rajesh Gopal (1992). Fundamentals of wildlife management, Justice Home, Allahabad, India

Sharma, P. D. (1985). Ecology and Environment. Rastogi Publications, Meerut.

Singh, H. R. (1990). Animal Ecology and Environmental Biology. Shobanlal Nagin Chand and Co. Delhi

Subrahmanyam, N. S. and Sambamurty, A. V. S. S. (2000). Ecology. Narosa Publishing House, New Delhi.

Subbarao, S. (2001). Ethics of Ecology and Environment. Rajat Publications, New Delhi.

The Wildlife (Protection) Act 1972. Natraj Publishers, Dehra Dun.

UNAIDS. (2004) Report on the Global AIDS epidemic, July 2004.

United Nations Department of Public Information DPI/1627/HR-March 1995.

Notes

Notes

Notes

Notes